Foundations of Organic C

MW01593063

1. Stereochemistry

The University of Liverpool
Department of Chemistry
Computer Assisted Learning Group

D. J. Chadwick
P. P. Duce
T. L. Gilchrist
D. Margerison
R. C. Storr
S. M. Walker

Longman
Scientific &
Technical

Copublished in the United States with
John Wiley & Sons, Inc., New York

Longman Scientific & Technical
Longman Group UK Limited,
Longman House,
Burnt Mill, Harlow,
Essex CM20 2JE, England
and Associated Companies throughout the world

Copublished in the United States with
John Wiley & Sons Inc., 605 Third Avenue, New York, NY 10158

© Longman Group UK Ltd 1989

First Published 1989

British Library Cataloguing in Publication Data

Foundations of organic chemistry - a software course. Stereochemistry
 1. Organic chemistry
 I. Chadwick, D. J.
 547

 ISBN 0-582-03567-8

Library of Congress Cataloguing in Publication Data available.

CONTENTS

How to use this book
Representation of three-dimensional structures

HOW TO USE THIS BOOK

This book is intended to accompany the four disks on Stereochemistry in the Foundations of Organic Chemistry series. Whilst the maximum benefit may be obtained by using the book and computer programs together, it has been written as a self contained introduction to Stereochemistry and may be used independently of the disks. The order in which the sections are arranged in this book is the same as on the disks. The book contains more detailed explanations of the terms and concepts on the disks, it extends them to cover material which is not on the disks, and it has extra problems (with answers at the back).

We suggest that you work through each of the tutorial disks first. If there are parts of the tutorial which you find difficult or do not understand, make a note of them and look them up in the appropriate section of the book afterwards. Even if you find the disks easy to follow, try to work through the book material after completing each disk, because you will find extra material in the book. For the Stereochemistry exercises on disk you will find it useful to have a set of molecular models to hand. You will also need to use the models for some of the exercises in this book.

REPRESENTATION OF THREE-DIMENSIONAL STRUCTURES

There are several different ways of representing three-dimensional structures on paper. On the disks and in this book, bonds coming forward, out
of the paper, are shown as bold wedges, and those directed behind the plane
of the paper are shown as dashed wedges, thus:

Other common ways of representing the same structure are as follows:

central carbon atom omitted

In more complex structures containing fused rings you may also see the
following conventional representation:

indicates H above plane at ring junction

equivalent to

indicates H below plane at ring junction

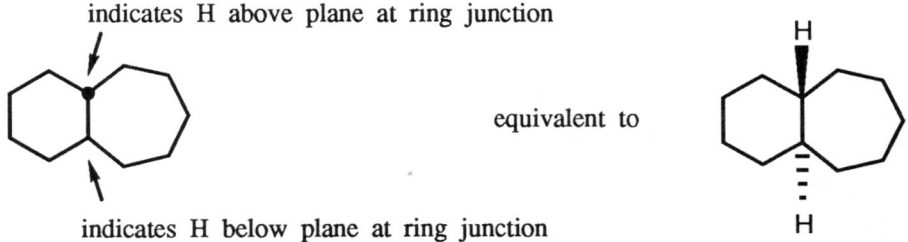

We have also sometimes used a shorthand representation of some functional groups in the structures on disk and in the book. These are as follows:

Me for CH_3
Et for CH_2CH_3
Pr for $CH_2CH_2CH_3$
Ph for C_6H_5 (phenyl).

STEREOCHEMISTRY

DISK 1: CHIRALITY. *R* AND *S* NOMENCLATURE

1.1 Chiral Molecules

The everyday objects which surround us can be classified into two broad groups. One type of object - for example, a ball, or a cube - appears the same when viewed from many different angles. Its reflection in a mirror is identical to the object itself. For the other type of object - for example, the wing of an insect, or our own hands and feet - this is not true. The "mirror image" of such an object cannot be fitted on top of the object itself; instead, the object and its reflection constitute a non-identical pair. Objects of this second type can be described as **CHIRAL**. "Chirality" is a term indicating "handedness"; that is, the non-identical relationship between an object and its mirror image. Objects of the first type, with mirror images identical to the objects themselves, are **ACHIRAL**.

The terms "chiral" and "achiral" can also be applied to molecules.

> **Exercise 1.1.** Make two models of bromochloromethane, CH_2BrCl. Arrange them as "mirror images" as shown below:

If you now pick up the right hand model you will find that it is the same as the left hand one (you can fit one model over the other). This shows that bromochloromethane is an ACHIRAL molecule.

Now, with the two models arranged as before, replace the top hydrogen atom on each one by a methyl group. You should have the two models shown:

These are models of 1-bromo-1-chloroethane. If you now try to fit the right hand model over the left hand one, you will find that the two models are not the same. 1-Bromo-1-chloroethane is a CHIRAL molecule.

1-Bromo-1-chloroethane is an example of the simplest type of chiral organic molecule: one which has four different substituents attached to a tetrahedral carbon atom. We describe such a carbon atom as **ASYMMETRIC**. The two non-identical mirror image forms are called **ENANTIOMERS**.

Exercise 1.2. Decide whether each of the following contains an asymmetric carbon atom, and so is chiral. Use models if you find it helpful.

(a)

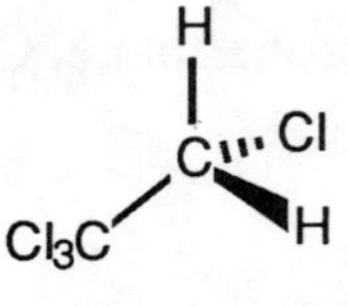

(b)

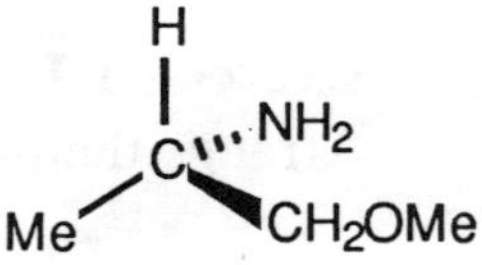

(c)

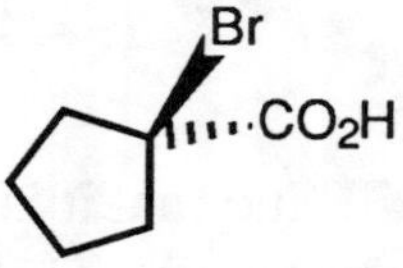

(d)

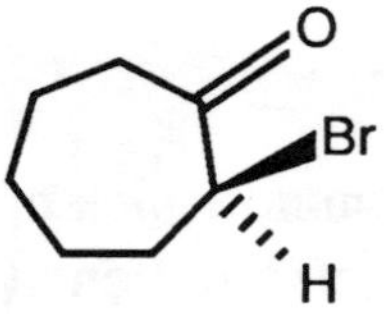

The two enantiomeric forms of 1-bromo-1-chloroethane are of equal energy. If we make this compound from an achiral precursor, and using achiral reagents, it will consist of equal numbers of the two enantiomeric molecules. We describe such a compound as a **RACEMATE** or a **RACEMIC MIXTURE**. If, on the other hand, we were able to devise a method of making the compound which gave only one of the two enantiomers, all the molecules of the product would be identical. Such a chiral substance consisting of only one enantiomer can be described as **HOMOCHIRAL**.

Let us take another look at the structures of bromochloromethane and of 1-bromo-1-chloroethane, in order to decide what makes one achiral and the other chiral. The important difference is related to the SYMMETRY of the two. In particular, bromochloromethane has a **PLANE OF SYMMETRY** (also called a MIRROR PLANE) which passes through the bromine, carbon, and chlorine atoms. By a "plane of symmetry" we mean a plane which divides the molecule into two identical halves. When we try to construct a similar plane in 1-bromo-1-chloroethane, by drawing it either through the same three atoms as before or anywhere else, we find that we can no longer do so. A chiral molecule does not possess a plane of symmetry.

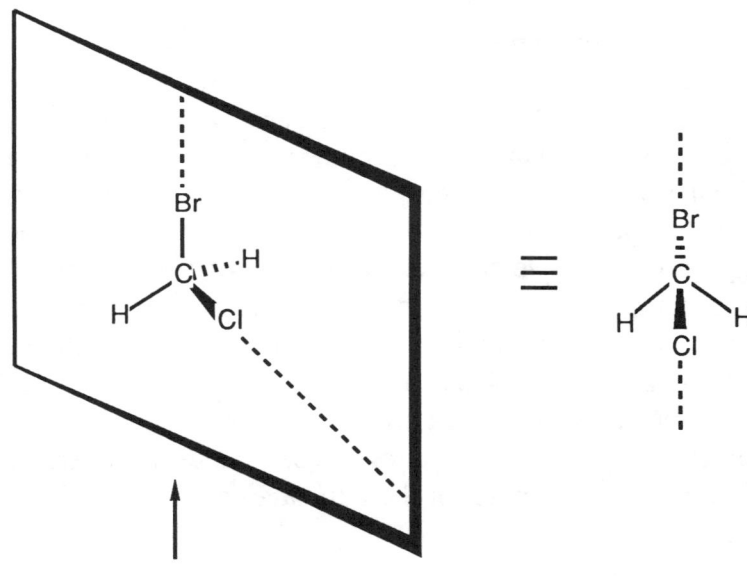

Plane of symmetry through Br, C and Cl   Plane of symmetry

Stereochemistry

Exercise 1.3. Go back to the structures in Exercise 1.2 which you decided were achiral, and draw a plane of symmetry through each.

It **is** possible for molecules to be achiral even though they lack a plane of symmetry. These types of achiral molecule are relatively uncommon in organic chemistry and they are not illustrated on the disks. Some such molecules have a **CENTRE OF SYMMETRY**. This is a unique point in the molecule such that on any straight line drawn through it, there is an identical environment in either direction at the same distance from the point. This is most easily understood by looking at examples. Two are shown below. Make models of them to convince yourself (i) that they are achiral and (ii) they have a centre of symmetry but not a plane of symmetry.

All the chiral molecules on the disks are chiral because they contain one or more asymmetric carbon atoms. There are, however, some other organic molecules which are chiral yet do not contain an asymmetric carbon atom. First, there are molecules which are chiral because of asymmetry about an atom other than carbon. Thus, a phosphine with three different substituents attached to phosphorus is chiral. The molecule is pyramidal and there is a high energy barrier to inversion, or "flipping", of the pyramid. It is interesting to compare this with a tertiary amine bearing three different substituents. This too is pyramidal and apparently chiral, but the energy barrier to inversion is so low (about 25 kJ mol^{-1}) that it is in rapid equilibrium with its enantiomer at room temperature. With a few exceptions, enantiomers of asymmetrically substituted nitrogen compounds are interconverted too rapidly at room temperature for them to have separate existence.

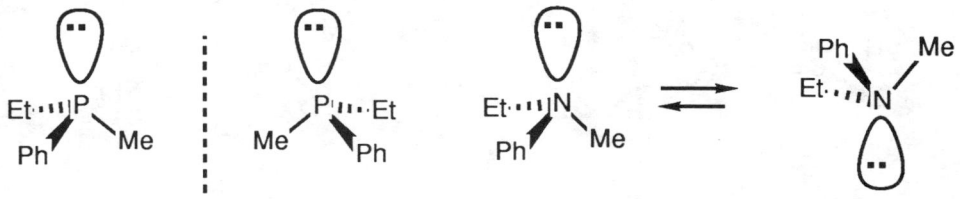

Some common sulphur compounds are also chiral. Sulphoxides with two different substituents attached to sulphur are examples. Like the phosphines, these molecules have a pyramidal structure with a high energy barrier to inversion.

A second group of molecules which are chiral but without asymmetric carbon contains rigid or bulky frameworks and substitution patterns which remove symmetry from the structures. Two examples are shown overleaf. The first is an allene. This class of molecule has two cumulated double bonds of which the central carbon atom is *sp* hybridized: the two π - lobes are thus at right angles. With substituents such as those shown the molecule has no plane of symmetry and is therefore chiral. The second example is a biphenyl derivative. Because this molecule has bulky substituents at the *ortho* positions it cannot be planar. It is chiral, and the enantiomers shown cannot interconvert at room temperature because the energy barrier to rotation about the central bond is too large. (Without bulky groups at the *ortho* positions the energy barrier to rotation is low enough to allow the enantiomers to interconvert at room temperature).

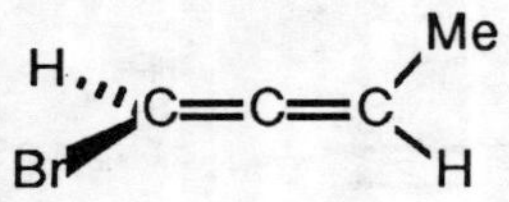

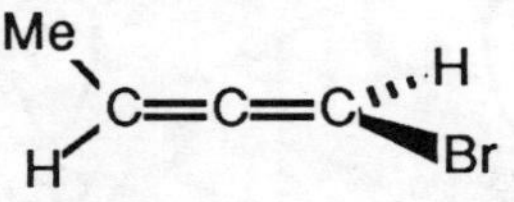

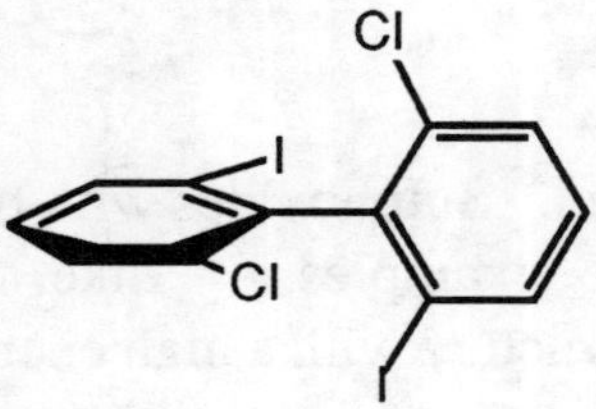

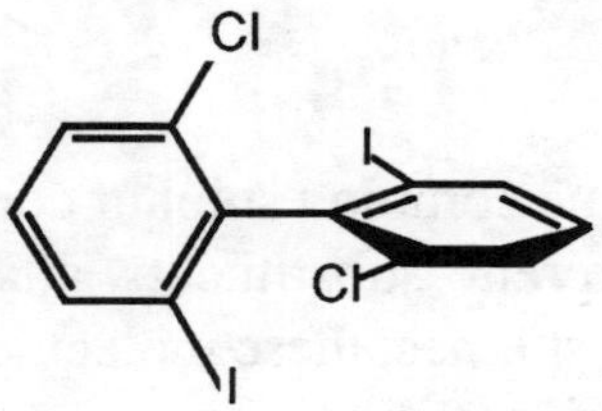

Exercise 1.4. Which of the following will exist in enantiomeric forms?

(a)

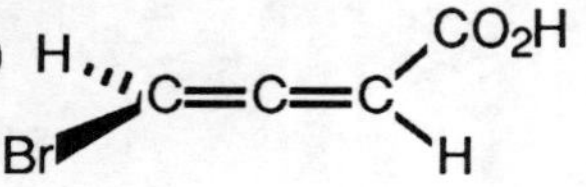

(b)

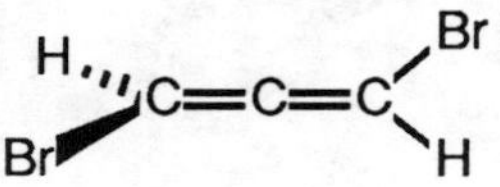

(c)

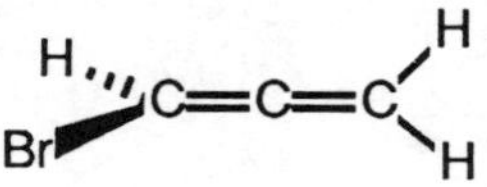

(d)

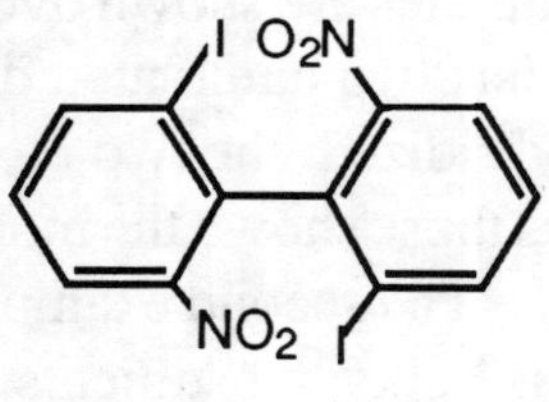

(e)

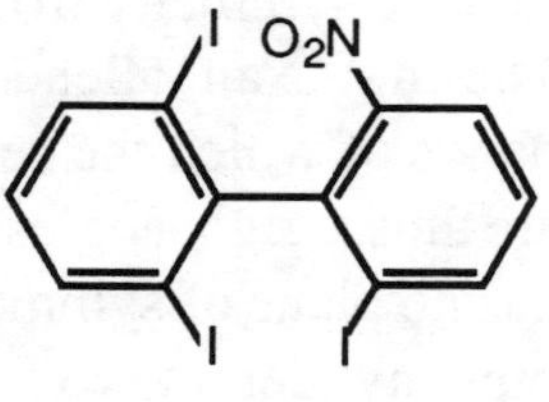

(f)

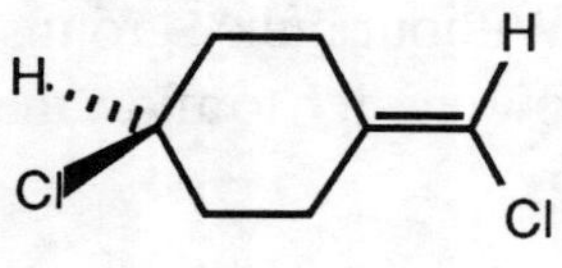

(g)

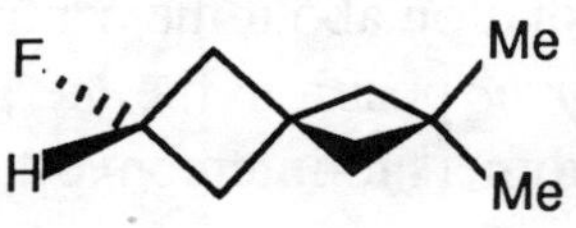

1.2 Optical Activity

You may wonder why it is important to be able to decide whether a particular substance can exist in enantiomeric forms or not. In most chemical reactions the enantiomers behave in an identical way. Most of their physical properties are also the same. There are, however, important exceptions to both of these generalizations. Reactions involving reagents which are themselves homochiral occur at different rates for the two enantiomers. For example, enzymes, which are large homochiral molecules, often react with the two enantiomeric forms of the same substance at greatly different rates (in many cases they effectively interact with only one of the two enantiomers). You can see why this should be so by imagining the active site of the enzyme as a chiral receptor "host" into which only one enantiomeric form of a chiral "guest" molecule can fit. Or, think of your right shoe as a chiral receptor and your feet as an enantiomeric pair of objects. The shoe differentiates nicely between this chiral pair!

The importance of this recognition of different enantiomers by enzymes is illustrated by the tragic history of thalidomide. This drug which was introduced in the 1950's as a sedative, has also been used to treat leprosy. It soon became clear that the drug was responsible for abnormalities in babies whose mothers had taken it during pregnancy, and it was withdrawn from use. The drug was marketed as a racemic mixture, but research in the late 70's established that the adverse effects were due entirely to one enantiomer, the S-(-)-form shown below. The other enantiomer has no such effects. Evidently the enzyme on which the drug acts to produce the damaging side effects has a chiral receptor site.

(S)-(-)- Thalidomide

An important case in which physical properties of enantiomers differ is in their interaction with plane polarized light. Chiral molecules possess the property of rotating the plane of polarized light when it is passed through a solution containing the substance. This property is called **OPTICAL ACTIVITY**. Enantiomers are sometimes referred to as **OPTICAL ISOMERS**.

Uneven interaction of the oscillating field of plane-polarised light with a chiral molecule. "Left is not equal to right"; the consequence is a rotation of the plane of polarisation.

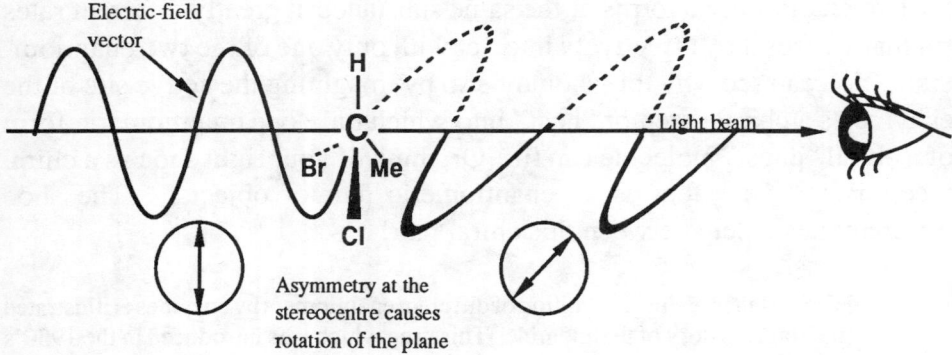

Optical activity can be detected, and measured, in the laboratory by an instrument called a polarimeter. A schematic form of a polarimeter is shown below.

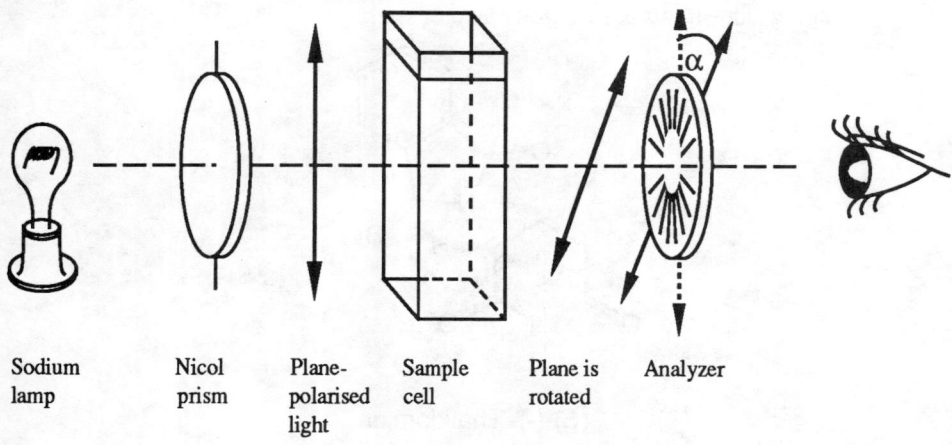

| Sodium lamp | Nicol prism | Plane-polarised light | Sample cell | Plane is rotated | Analyzer |

The polarimeter allows the degree of optical rotation, α, of the plane of the polarized light to be measured. Obviously this measured value depends on a number of variables including the concentration of the solution containing the substance, the path length of the solution (that is, the length of the solution cell), and the temperature. If each of these factors is corrected to a standard then a standard value of α, called the **SPECIFIC ROTATION**, can be determined. **The specific rotation is a constant for a single enantiomer of a particular substance.** Normally, polarimeters operate using light of a single wavelength (589 nm) from a sodium source. This monochromatic source is the sodium D line. The specific rotation at a temperature t°C is represented by $[\alpha]^t_D$ and is related to the measured value by the expression

$$[\alpha]^t_D = \frac{\alpha}{l\,c}$$

where l is the path length in decimetres (1 decimetre = 10 cm) and c is the concentration of the solution in $g\ cm^{-3}$.

Exercise 1.5. Calculate the specific rotation at 20°C of a solution in ethanol of a substance for which the measured rotation α = +1.70° (at 20°C) in a 5 cm cell at a concentration of 8.0 g per 100 cm³ solution.

You may notice that in the exercise above the sign of the observed rotation is quoted. Some homochiral substances rotate the plane of polarised light to the **right**; that is, they are **DEXTROROTATORY**. Solutions of these substances have positive values of α. Others rotate the plane to the **left**; that is, they are **LAEVOROTATORY** and have negative values of α. The direction of rotation is a purely experimental observation and cannot be predicted from structure in any simple way. You may see the direction of rotation of a particular homochiral substance quoted with its name; e.g., (+)-tartaric acid.

Stereochemistry

Another way of indicating the direction of rotation which was once widely used, but which is no longer recommended, is by means of prefixes d and l. The symbol d (for dextrorotatory) indicates a positive value of α, and l (for laevorotatory), a negative value.

Achiral substances have $[\alpha]_D$ values of zero. As we have seen, chiral substances which contain molecules of a single enantiomer normally have measurable, positive or negative, values of $[\alpha]_D$. **Opposite enantiomers of the same substance have values of $[\alpha]_D$ which are equal but of opposite sign** at any given temperature. It follows from this that a racemate, containing equal numbers of molecules of each enantiomer, will not rotate the plane of polarized light: the contributions from the two sets of molecules cancel each other out. This means that a measurable $[\alpha]_D$ indicates the presence of at least a partial excess of one enantiomer whereas a zero $[\alpha]_D$ can indicate the presence of either an achiral substance or a chiral substance in its racemic form.

Now consider how we can interpret an observed $[\alpha]_D$ value when one enantiomer is present in excess but some of the opposite enantiomer is also present. This is a common situation when asymmetric synthesis is carried out in the laboratory! For example, the reduction of an unsymmetrical ketone such as butan-2-one by an **achiral** reducing agent such as sodium borohydride leads to the formation of a **racemic** alcohol. However there are now also several **homochiral** reducing agents available for the conversion of ketones into secondary alcohols. When these are used on an unsymmetrical ketone such as butan-2-one, the product butan-2-ol consists of a mixture of both enantiomers **but in unequal amounts**. The usefulness of the reducing agent depends upon how successful it is in producing one enantiomer selectively.

(unequal amounts)

We can express the relative amounts of the two isomers in terms of the **OPTICAL PURITY** of the substance. [This is sometimes also referred to as ENANTIOMERIC EXCESS (e.e.)]. There are now reliable experimental methods for determining optical purity. (Most are based on n.m.r. spectroscopy and use homochiral solvents, complexing agents, or reagents to distinguish between the enantiomers). If we know the $[\alpha]_D$ value for a single enantiomer then it is also a simple matter to relate optical purity and the experimentally observed $[\alpha]_D$.

As an example, consider a sample of a substance which consists of 80% of the molecules of the (+)-enantiomer and 20% of the molecules of the (-)- enantiomer. If $[\alpha]_D$ for the pure (+)-enantiomer is +100°, what value should we expect for the sample?

The diagram below shows that one in every four of the molecules of the (+)-enantiomer cancels out the contribution from the molecules of the (-)- enantiomer. This leaves three molecules out of every five to contribute to the observed rotation, which is therefore +60°. The optical purity of the substance is defined as follows:

% Optical purity = observed $[\alpha]$ / $[\alpha]$ for single enantiomer x 100

In this example, therefore, the optical purity is 60% (note that it is **not** the same as the percentage of the major enantiomer present).

Stereochemistry

Exercise 1.6. Calculate (a) the optical purity and (b) the ratio of the two enantiomers for a sample of 2-bromobutane with an observed $[\alpha]_D$ at 20°C of -20.0°. Assume that $[\alpha]_D$ of pure (-)-2-bromobutane is -25.0° at 20°C.

1.3 Absolute Configuration: *R* and *S* Nomenclature

We saw in Section 1.2 that enantiomers can be distinguished by the sign of their experimentally determined $[\alpha]_D$ values. Unfortunately, this does not tell us which enantiomer is represented by which structure. We need to use some technique such as X-ray crystallography to determine the structure of a particular enantiomer. The correct three-dimensional arrangement of the substituents about the asymmetric carbon atom is called the **ABSOLUTE CONFIGURATION**.

If we want to describe a particular enantiomer we need a system of nomenclature which allows us to specify absolute configuration. An unambiguous way of doing this was devised by three chemists, R. S. Cahn, C. K. Ingold, and V. Prelog. The **Cahn-Ingold-Prelog** system is now the commonly accepted way of indicating absolute configuration.

An earlier method, which you may still see in use, was to link the absolute configuration of a given substance to one of known absolute configuration through a series of chemical reactions for which the mechanism was known. The two enantiomers of 2,3-dihydroxypropanal (glyceraldehyde) were labelled as D and L. They were assigned the absolute configurations shown. At the time this was done, there was no way of knowing whether this assignment was valid, but it was later established experimentally that the guess had been correct. Any other molecules which could be shown chemically to be related to either of these, and thus to have analogous stereochemistry about the central carbon atom, were assigned to the D or the L series. For example, the acid formed by oxidation of D-glyceraldehyde was assigned as the D-isomer, with the configuration shown, since it was assumed that the stereochemistry about the asymmetric carbon atom did not alter in this reaction. (Note that this acid has a negative sign of rotation, whereas D-glyceraldehyde has a positive value: there is no relationship between the label D or L and the experimentally observed sign of rotation). The method is not applicable to all chiral molecules, however, and it is being superseded by the Cahn-Ingold-Prelog system.

D-(+)-Glyceraldehyde   L-(-)-Glyceraldehyde   D-(-)-Glyceric acid

The Cahn-Ingold-Prelog system is based on the ranking of substituents in terms of priorities. Like other systems of nomenclature, this is just a convention which everyone agrees on. The substituents attached to an asymmetric carbon atom are then labelled in order of decreasing priority.

The first example on the disk shows bromochlorofluoromethane. The four different atoms attached to carbon are ranked 1 to 4 in the order of decreasing atomic number: Br>Cl>F>H.

The general procedure for naming the enantiomers, which is illustrated for bromochlorofluoromethane, is as follows.

1 Rank the groups in order of decreasing priority and view the structure with the group of lowest priority (hydrogen in this case) pointing directly away from the line of vision.

2 Rank the three remaining groups, from the direction of view, in order of decreasing priority (in this case, Br, Cl, and F). Draw a curved arrow linking these three groups.

3 If these groups occur in a clockwise direction, the isomer is given the label R. If they occur in an anticlockwise direction, the isomer is given the label S. [The labels R and S are derived from the Latin rectus (right) and sinister (left)].

Stereochemistry

The procedure is illustrated below for the enantiomers of bromochlorofluoromethane.

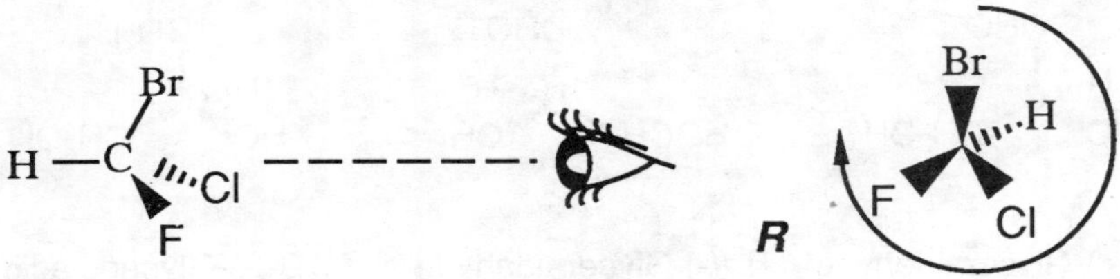

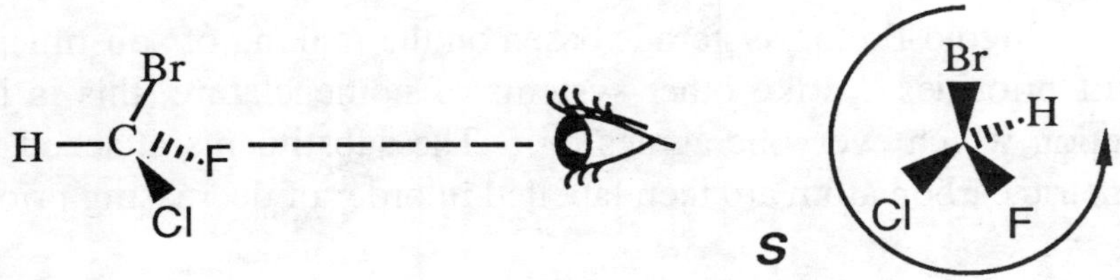

It is easy to decide the order of priority of groups which are single atoms: all we need to know is the atomic number of the atom concerned. More commonly, substituents are made up of several atoms. When you are deciding on the relative priorities of such groups, a general guideline is NOT to work out the total relative mass of the atoms in the group, but to look instead at the order of attachment of atoms to the asymmetric carbon centre.

THE PRIORITY RULES

(a) Substituents are ranked in the order of decreasing atomic number of the atoms attached directly to the asymmetric centre.

Example: NH_2 ranks higher than CCl_3.

(b) When isotopes of the same element are present, the atoms are ranked in order of decreasing isotopic mass.

Example: D (deuterium, or ^2H) ranks higher than H.

(c) If two atoms attached to the asymmetric centre are identical the next atoms along the substituent chain are used for priority assignment. If these atoms also have identical atoms attached to them, priority is determined at the first point of difference along the chain. The atom that has attached to it an atom of higher priority has the higher priority.

Examples: CH_2Cl ranks higher than CH_2CH_3.
CH_2OH ranks higher than $CH(CH_3)_2$.
$CH(CH_3)_2$ ranks higher than CH_2CH_2OH.

(d) Where an atom is attached to another by a double bond or a triple bond it is treated as having two, or three, single bonds of the same type. Priority rules are then assigned as in (a) to (c) above.

Examples: $CH=CH_2$ is treated as $CH-CH_2-C$
$$|$$
$$CH_2-C$$

CO_2H is treated as $HO-C-O-C$
$$|$$
$$O-C$$

CN is treated as $C-N-C-N-C$
$$|$$
$$N-C$$

Thus CHO has higher priority than CH_2OH.
$CH=CH_2$ has higher priority than CH_2CH_2Cl.

Exercise 1.7. Choose from each of the following pairs the substituent of higher priority.

(a) OH and CBr_3.
(b) $CHCl_2$ and $C(CH_3)_3$.

(c) C₆H₅ (phenyl) and CH₂NH₂.

(c) C_6H_5 (phenyl) and CH_2NH_2.

(d) $CONH_2$ and CN.

(e) $CH=CH_2$ and $CH(CH_3)_2$.

Exercise 1.8. Rank the following sets of substituents in order of decreasing priority.

(a) CH_3, CH_2D, F, and Li.

(b) C_6H_5, $CH=CH_2$, CN, and CHO.

(c) CH_2CH_2OH, $CH=CH_2$, $C\equiv CH$, and CH_2CF_3.

Once the priorities have been determined it is a straightforward matter to determine whether the absolute configuration is R or S. In many cases it is necessary to rotate the molecule in order to put the group of lowest priority at the rear. Try the following exercises: use models if you find it difficult to rotate the molecule in order to view it from the correct direction.

Exercise 1.9. Assign each of the following as R or S.

Exercise 1.10. Draw structures for each of the following:
(a) (*R*)-butan-2-ol
(b) (*S*)-2-chloropropanoic acid
(c) (*S*)-2-methylbut-3-ynal
(d) (*R*)-2-chlorocyclohexanone.

The final exercise on Disk 1 requires you to identify, and assign the absolute configuration of, all the asymmetric carbon atoms in morphine. You will be getting more practice in assigning stereochemistry to molecules with more than one asymmetric centre on Disk 3. If, however, you tried the problem and want to know the solution, here it is:

C-5 is *R* : 1, O; 2, C-6; 3, C-13; 4, H
C-6 is *S* : 1, OH; 2, C-5; 3, C-7; 4, H
C-9 is *R* : 1, N; 2, C-14; 3, C-10; 4, H
C-13 is *S* : 1, C-5; 2, C-12; 3, C-14; 4, C-15
C-14 is *R* : 1, C-9; 2, C-13; 3, C-8; 4, H

(Note that nitrogen is not chiral because of rapid inversion).

Stereochemistry

DISK 2: PROJECTIONS

We saw on Disk 1 that three-dimensional structure can be represented
unambiguously by the use of wedged and dashed bonds. This type of
structural formula is the one in most common use and it is the best for most
purposes. There are several other conventional methods which you may
come across for the representation of stereochemistry in formulae. The aims
of Disk 2 are to introduce you to three types of these so-called
PROJECTION FORMULAE and to show how they are related to the
more common "dash and wedge" formulae.

2.1 Sawhorse Projections

Sawhorse projections are simplified line formulae which are used for small
open chain carbon compounds.   The sawhorse representation of such a
compound is compared with its dash-wedge structure below.

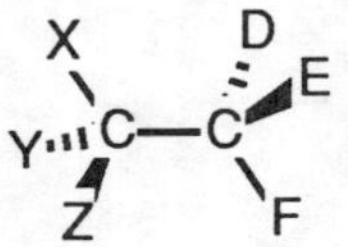

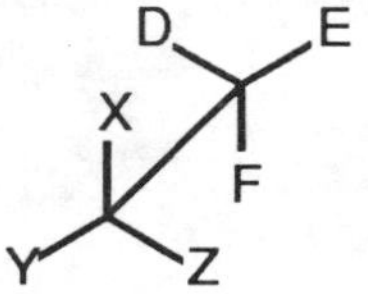

The sawhorse representation has the advantage of clarifying the nature and
relative position of the substituents.   It is quick and easy to draw by hand.

 When the carbon atoms are asymmetrically substituted you need to be sure
which is the "front" carbon atom of the two in the projection formula.   The
usual convention is that it is the lower left hand atom, as shown on the
opposite page.

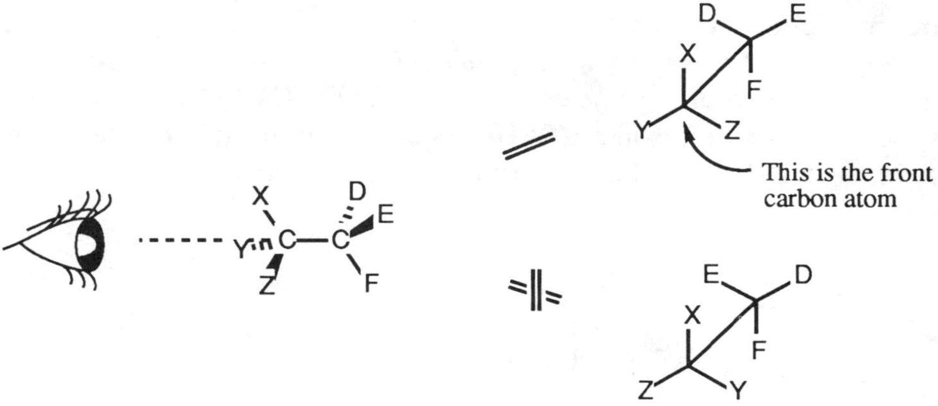

This is the front carbon atom

Exercise 2.1. Draw sawhorse projection formulae for each of the following:

(a)

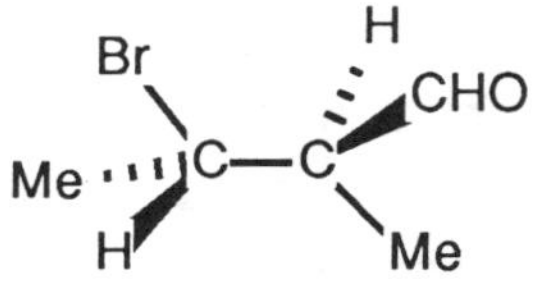

(b)

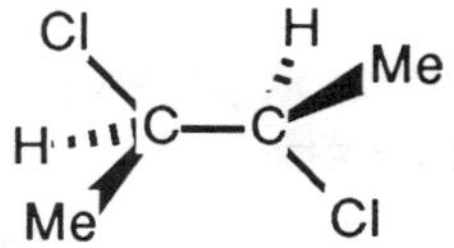

Exercise 2.2. Draw dash-wedge structures for each of the following

(a)

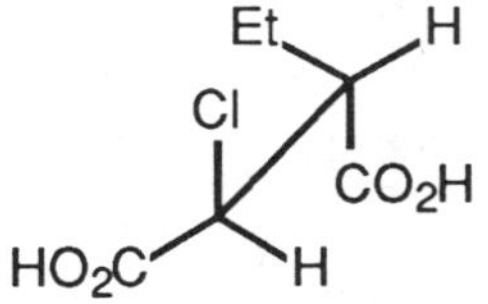

(b)

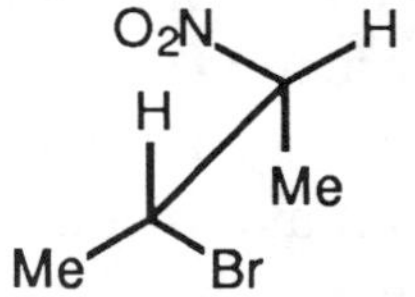

In all of the examples illustrated above, the substituents on the two carbon atoms are arranged so as to place the groups on the second atom as far away as possible from those on the first atom. You will notice from the sawhorse projections that the substituents on one carbon atom bisect the angle between the substituents on the other. A different arrangement is shown below, in

Stereochemistry

which the substituents on the two atoms are directly aligned. Since there is free rotation about the central bond of such simple acyclic compounds, these do not exist independently but are rapidly interconverted at room temperature. The two arrangements are called **CONFORMATIONS**. The structure on the left is in a **STAGGERED** conformation and that on the right, in an **ECLIPSED** conformation.

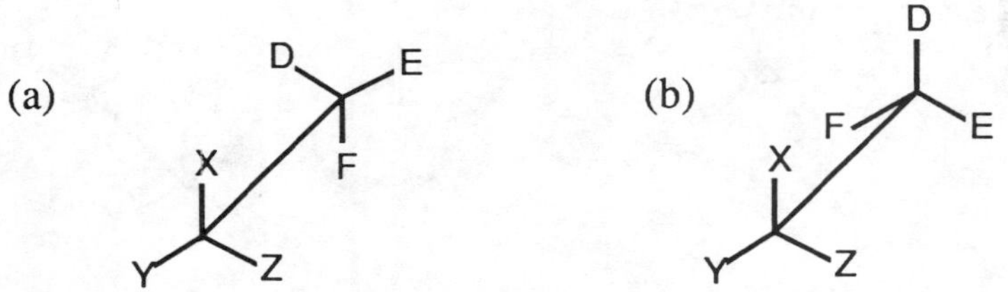

(a) Staggered and (b) eclipsed conformations.

The interconversion of such conformations is illustrated and discussed on Disk 4. Note, however, that sawhorse projection formulae are quite useful for illustrating the different conformations. Another type of projection formula which is also commonly used to represent such conformations is described in the next Section.

2.2 Newman Projections

A **NEWMAN PROJECTION** is an easy-to-draw shorthand method of representing the relative positions of substituents on adjacent carbon atoms. If you imagine looking along an axis represented by the bond joining the two carbon atoms, then the **rear** carbon atom is represented in the Newman projection by a circle, with the three substituents attached at appropriate points to the circumference, thus:

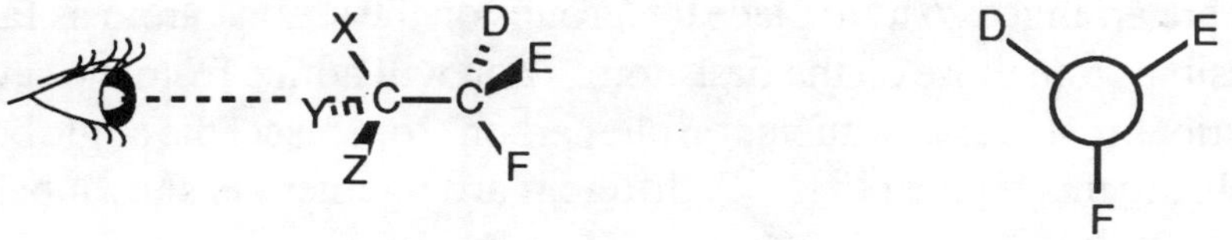

The **front** carbon is represented by a point at the centre of the circle, its three substituents being joined by bonds to this central point. You can compare dash-wedge, sawhorse, and Newman representations of chloroethane (in a staggered conformation) below.

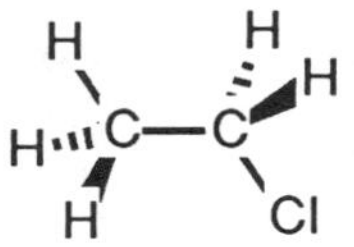

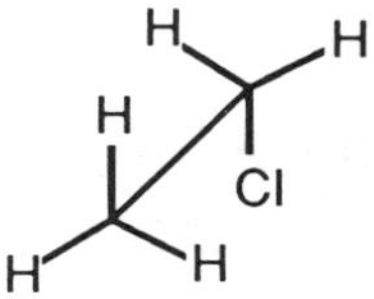

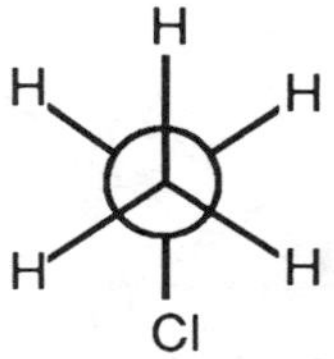

One advantage of the Newman projection formula is that it is easy to see the angle between the substituents on the two carbon atoms. In the staggered conformation shown, this is 60°. If the substituents on the rear carbon atom are rotated through 60°, the structure has an eclipsed conformation. The eclipsed conformation is represented by the Newman projection shown below, the rear substituents being drawn slightly "off-line" in order to make them visible.

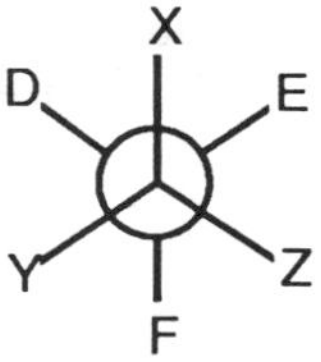

staggered 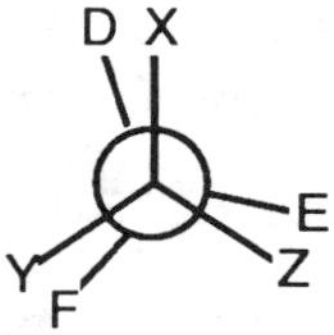   eclipsed

Exercise 2.3. Draw Newman projections for each of the following:

(a) (b) 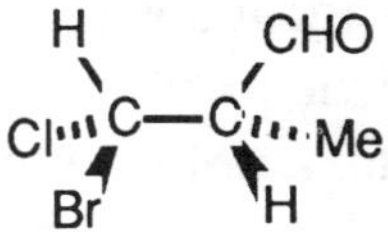 (c)

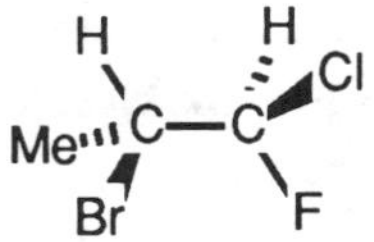

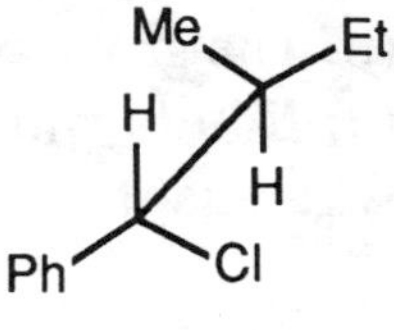

Stereochemistry

Exercise 2.4. Assign the absolute configuration of the front carbon atom of the following as R or S.

(a)

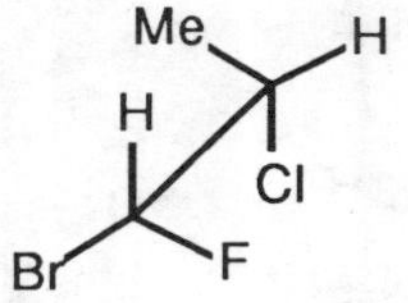

(b)

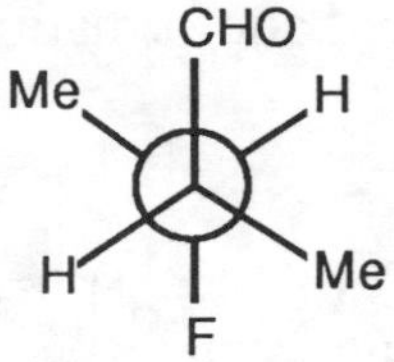

2.3 Fischer Projections

The first type of projection formula to be widely used was introduced by the German chemist Emil Fischer in order to represent the configurations of the carbon atoms in sugar molecules. They are still sometimes used to represent the structures of carbohydrates and the convention by which the structures are drawn can be applied to the substituents about any tetrahedral carbon atom. These **FISCHER PROJECTION** formulae need to be interpreted with care: they appear deceptively simple!

Exercise 2.5. Make a model of (R)-2-chloropropanal and align it as shown below. Compare it with the Fischer projection formula for (R)-2-chloropropanal which is shown alongside.

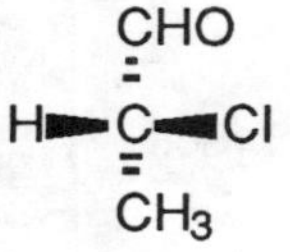

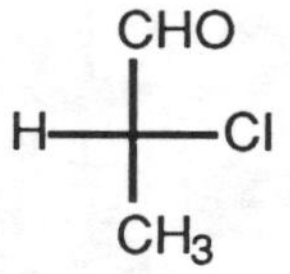

This example illustrates the features of the convention used for constructing these projection formulae, namely:
(a) The molecule is aligned with the longest carbon chain vertical.
(b) The group of higher priority at the end of this chain is placed at the top.
(c) **Bonds which are vertical project behind the plane of the paper, whereas those which are horizontal project in front of the plane.**

This last point is the most important one (Fischer projection formulae are sometimes drawn without strict observance of the other two points). This is the point which is also the most easily misinterpreted. Consider, for example, the result of rotating the model of 2-chloropropanal through 90°. This is **not** the same structure as the one we arrive at by rotating the Fischer projection formula through 90°, as the drawings below illustrate (it is in fact the enantiomer). In any Fischer projection formula the vertical bonds are behind the plane of the paper.

Exercise 2.6. Rotate the Fischer projection formula corresponding to (R)-2-chloropropanal through 180°. Does this new formula represent the original molecule or not?

If the projection formula represents a molecule for which the stereochemistry is specified at more than one carbon atom, you will notice that it shows the molecule in an eclipsed conformation. Look at the general example overleaf (which is also on the disk). The pairs of substituents D and X, F and Y, and E and Z are aligned.

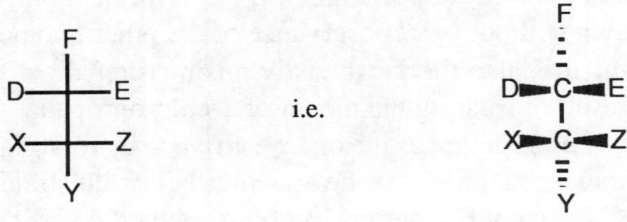

i.e.

The final part of Disk 2 provides practice in the use and interpretation of Fischer projections. Some additional exercises are given below.

Exercise 2.7. Draw Fischer projection formulae corresponding to each of the following. Take care to draw formulae with the longest carbon chain vertical.

(a)

Me CO_2H

H''C—C''' Cl

HO   H

(b)

HO_2C H

H''C—C''' CH_2OH

HO   OH

(c)

H   CHO

H

Me

Me   Et

(d)

CO_2H

HO   Me

MeCONH   H

H

Exercise 2.8. Label the asymmetric centres in the following Fischer projection formulae as *R* or *S*.

(a)

CO_2H

H_2N——H
     1

H——OH
     2

Me

(b)

CHO

HO——Me
     1

HO——H
     2

CH_2OH

DISK 3:   DIASTEREOMERS

3.1   Stereoisomerism in Molecules with Two Asymmetric Carbon Atoms

There are several structures on Disks 1 and 2 which contain more than one asymmetric carbon atom.   Structures of this type can exist in enantiomeric forms because of the presence of asymmetric carbon centres.   In addition, other stereoisomeric structures are possible which are not enantiomers.

An example of a compound which has a structure of this type is 2-chloro-3-phenylbutane, which is illustrated on Disk 3.   Structure (i) shows one stereoisomer, the 2S,3R isomer.   If we compare this with the three other stereoisomeric forms (ii) to (iv) shown, we can see that structure (ii) is an enantiomer of structure (i) but that the others are not.

Exercise 3.1. Make a model of structure (i) and compare it with structures (ii) to (iv) above. Convince yourself that none of these structures is identical to structure (i) and that only structure (ii) is a mirror image form of it.

Thus, if we compare structures (i) and (iii) we find that they are stereoisomers but not enantiomers.   Stereoisomers which are not mirror image forms are called DIASTEREOISOMERS.   The abbreviated form **DIASTEREOMERS** is commonly used.

Exercise 3.2. Identify, from structures (i) to (iv) on the previous page, another pair of enantiomers and any other two structures which are diastereomers.

Unlike enantiomers, diastereomers usually have quite different physical and chemical properties. They have different melting points, boiling points, and other physical constants. Their spectra (for example, n.m.r. spectra) are not superimposable. Reactions occur at different rates, often leading to different products.

An example of a structure which occurs naturally in different diastereomeric forms is 2-methylamino-1-phenylpropanol. Both the $1S,2R$ and the $1S,2S$ diastereomers are found in Nature. The two compounds have different melting points and other physical properties. Both compounds are used in medicine: their properties, although similar, have subtle differences and their most common uses are different. The $1S,2R$ isomer is (-)-ephedrine; and it is used as a bronchodilator ("Ephedral").

(-)-ephedrine   (+)-pseudoephedrine

The $1S,2S$ isomer is (+)-pseudoephedrine, which is commonly used as a decongestant ("Sudafed").

On Disk 3 there are exercises to provide practice in the identification of identical, enantiomeric, and diastereomeric structures. If you find these difficult, here are some more examples, which also provide further practice in the interpretation of projection formulae:

Exercise 3.3. For each of the following pairs of structures, decide whether they are identical, enantiomeric, or diastereomeric. (Assume that there is free rotation about the single bonds).

(a)

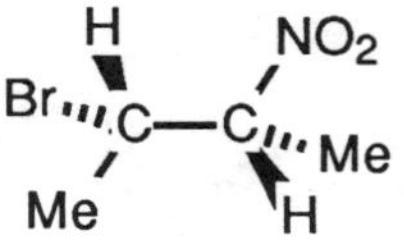

and

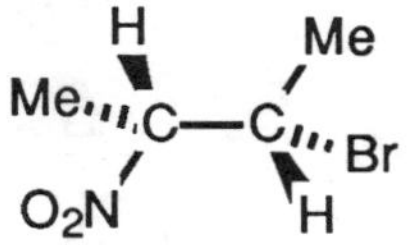

(b)

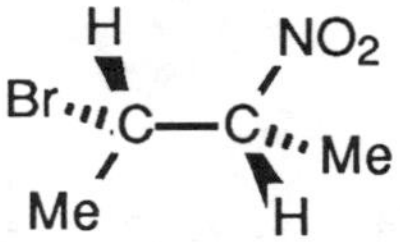

and

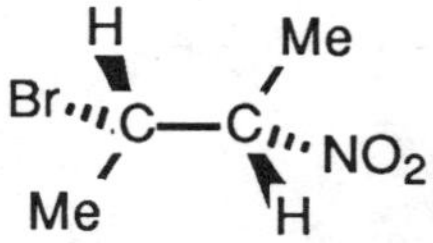

(c)

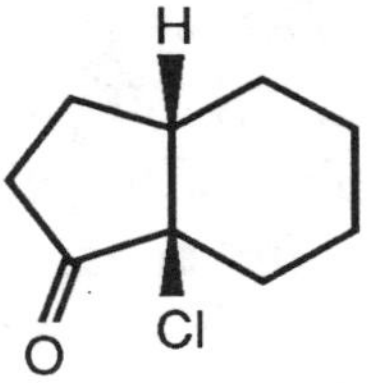

and

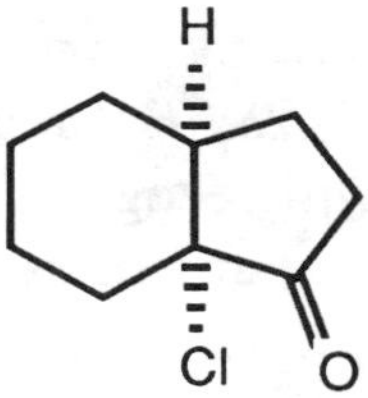

(d)

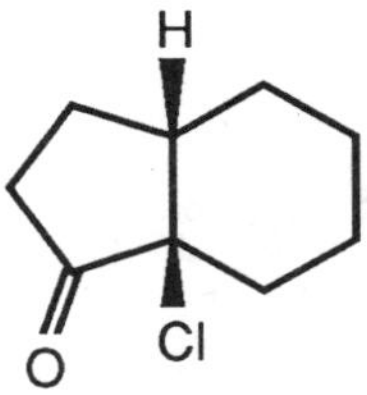

and

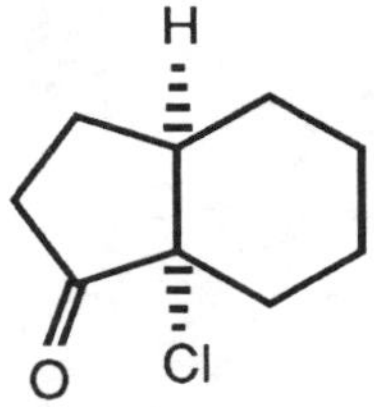

(e)

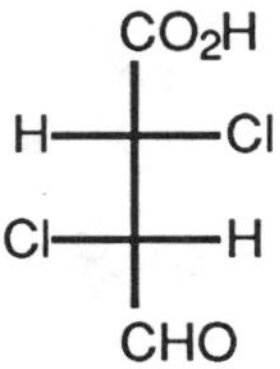

and

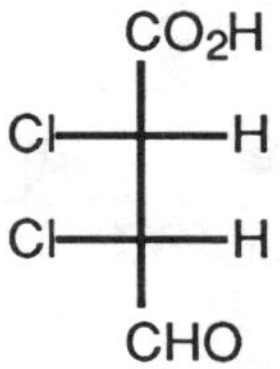

(f)

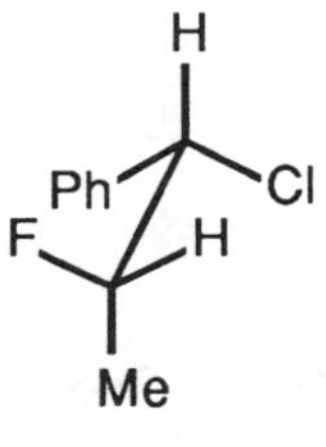

and

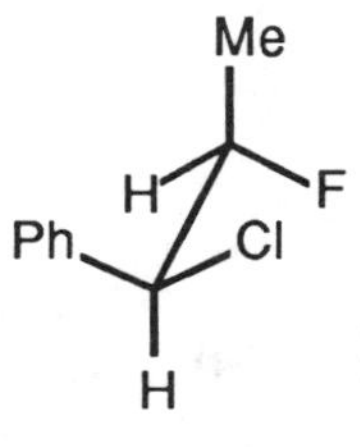

Stereochemistry

(g)

and

Note that 2-chloro-3-phenylbutane, which has two asymmetric carbon atoms, has four stereoisomers, which were illustrated earlier. We can identify these by the the absolute configurations of carbon atoms 2 and 3. They are (i) 2S,3R; (ii) 2R,3S; (iii) 2R,3R; (iv) 2S,3S. If you did Exercise 3.2 you will recognise that these four isomers consist of two pairs of enantiomers, structures (i) and (ii) and structures (iii) and (iv). This is also illustrated on Disk 3 with the example of 2-bromo-1-chloro-1-phenyl-propane. The four stereoisomers are:

- - - -   enantiomers   ᔟᔟ   diastereomers

We can deduce that **a structure containing two asymmetric carbon atoms can have a maximum of four stereoisomers.**

3.2. Structures With More Than Two Asymmetric Centres. The 2^n Rule.

If a structure contains more than two asymmetric carbon atoms the number of stereoisomers is obviously more than four. In fact there is a simple formula which we can use to work out the number. This is based on the number of possible ways of combining centres of a specified absolute configuration.

Consider a structure containing three asymmetric centres. Each can be designated as R or S. The ways in which three such centres can be combined are: R,R,R, R,R,S, R,S,R, S,R,R, S,S,R, S,R,S, R,S,S, and S,S,S: that is, eight, or 2^3, in all. Any additional asymmetric centre doubles the number of possible stereoisomers. Thus, **the maximum number of stereoisomers for a compound with n asymmetric carbon atoms is 2^n.**

Disk 3 contains several examples which provide practice in identifying asymmetric carbon atoms and then in applying the 2^n rule to calculate the number of possible stereoisomers. Three of these are well known natural products: Penicillin G, Vitamin D_2, and androsterone. Although the 2^n rule predicts a large number of possible stereoisomers for each, the natural products are single stereoisomers. They have the structures shown below (with the chiral centres identified by asterisks). The structure of morphine on Disk 1 provides another example in which many stereoisomers are possible, only one of which is the naturally occurring compound.

Penicillin G

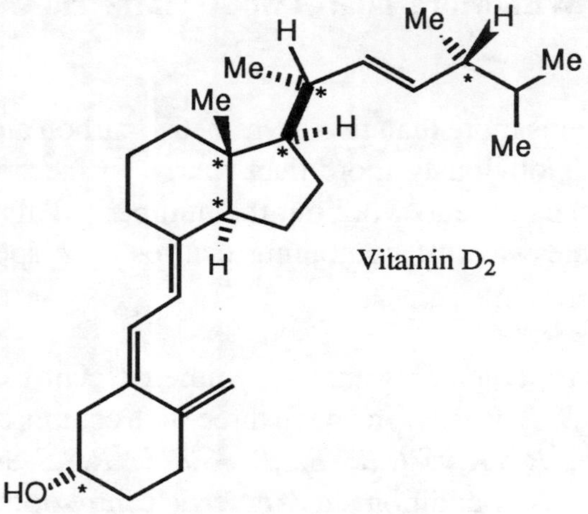

Vitamin D₂

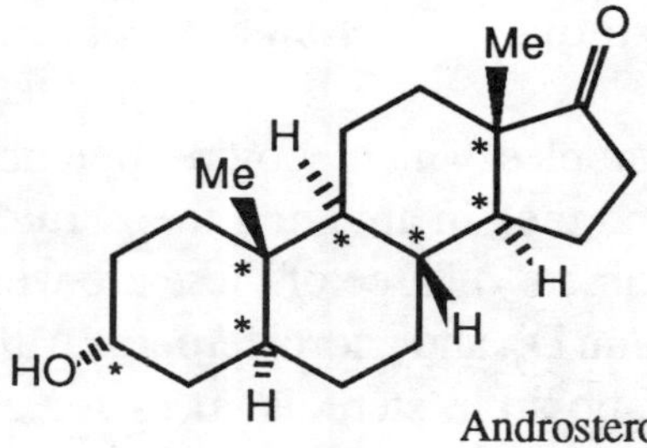

Androsterone

In order to understand the chemistry of compounds of this type we need to know more than just the carbon skeleton and the types of functional groups present: it is important to know the three dimensional structure of the particular stereoisomer concerned. This three dimensional structure can be quite different for different diastereomers. The following exercise, on a structure with only two asymmetric carbon atoms, illustrates the point. With more than two such centres the differences can be much greater.

Exercise 3.4. Make models of the two diastereomers illustrated opposite. Compare the models and note the differences, particularly in terms of rigidity and in the environment of the carbonyl group.

If you have the models available you will also find it useful to make a model of Penicillin G. From this you will be able to see how a change in configuration at any of the asymmetric centres substantially changes the overall shape of the molecule.

3.3. *Meso* Structures

It is possible for a structure to contain asymmetric carbon atoms and yet be achiral. This is the case when two asymmetric centres have the same sets of substituents and there is an internal plane of symmetry. Such structures are called **MESO** forms.

An example of a compound which has a *meso* structure is *cis*-cyclobutane-1,2-diol. This structure has two asymmetric carbon atoms (that is, they have four different substituents) but the sets of substituents are the same for the two atoms. Their arrangement in space is such that the molecule has an internal plane of symmetry, as shown. The structure is therefore achiral.

For *trans*-cyclobutane-1,2-diol the sets of substituents on the two asymmetric centres are also identical but the structure does not have an internal mirror plane. It is therefore chiral and exists in enantiomeric forms, as shown below.

There are thus three, not four, stereoisomers of cyclobutane-1,2-diol. The existence of a *meso* form reduces the number of isomers.

cis-Cyclobutane-1,2-diol is a fairly rigid molecule and the internal mirror plane is easy to recognise. In the case of *meso*-tartaric acid, which is illustrated on Disk 3, we have to draw it with the substituents eclipsed in order to see the internal mirror plane. This structure, too, is achiral. Tartaric acid has a total of three stereoisomers, the *meso* form being diastereomeric with each of the two chiral forms. You will notice that Fischer projection formulae are useful for emphasising the differences between these structures.

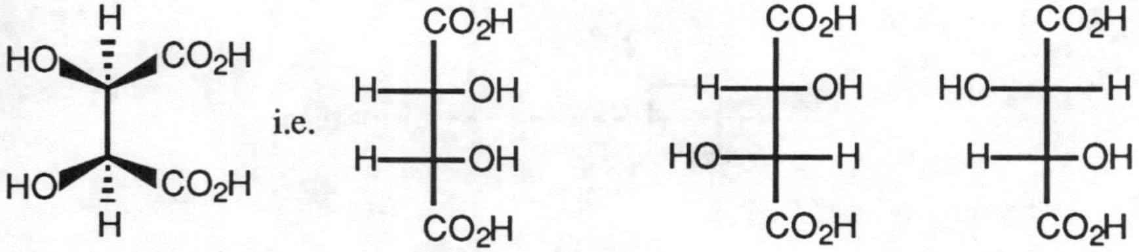

meso form (achiral)   enantiomeric forms

Exercise 3.5. Draw the *meso* isomers of each of the following structures and indicate the position of the internal mirror plane.

(a) MeCHClCH₂CHClMe

(b)

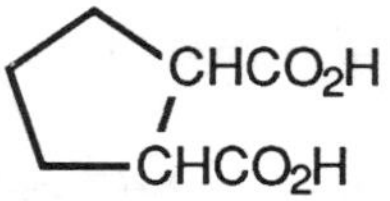

(c)

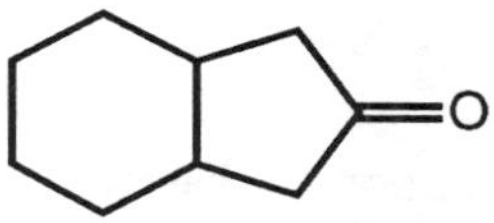

3.4 *Threo* and *Erythro* Nomenclature

The final section of Disk 3 provides an explanation of the terms *threo* structures and *erythro* structures which you may sometimes see applied to diastereomers. The terms are used for diastereomers with two asymmetric carbon atoms and with similar sets of substituents on the two carbon centres. The structures are arranged in an eclipsed conformation, in which any identical groups are aligned. Consider the example of 2-bromo-3-chloro-butanedioic acid, which is also shown on Disk 3. This has two asymmetric carbon atoms and four stereoisomers, which are illustrated by the four Fischer projection formulae below.

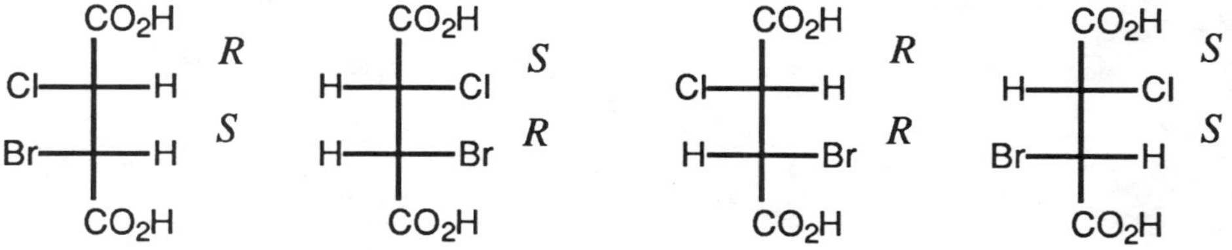

erythro   *threo*

In two of these structures, the 2*R*,3*S* and the 2*S*,3*R* isomers, the Fischer projection formulae show the maximum number of identical groups aligned with each other on the two carbon atoms. These are *erythro* forms. The other two isomers have only one pair of identical substituents eclipsed. These are the *threo* forms. You can see that the two *erythro* forms are enantiomeric and the term *erythro* can be applied to a racemic mixture of the two enantiomers. Similarly, the two *threo* isomers are enantiomeric.

Exercise 3.6. Draw *erythro* and *threo* Fischer projection formulae for (i) 2- chloro-3-iodobutane and (ii) 2,3-dibromopentane.

The investigation of reaction mechanisms has often depended upon the ability of chemists to differentiatebetween diastereomers of this kind. An illustration of this is on Disk 3. This shows the stereochemical outcome of the addition of bromine in methanol to a pair of alkenes, dimethyl fumarate [(E)-dimethyl butenedioate] and dimethyl maleate [(Z)-dimethyl butenedioate]. The reactions of the two alkenes lead to the formation of different diastereomers of dimethyl 2-bromo-3-methoxybu-tanedioate. As the structures on the following two pages and on the disk show, this can be explained by the formation and stereospecific ring opening of different bromonium ions from the two alkenes. Structures (1) and (4) are enantiomeric *erythro* isomers and structures (2) and (3) are enantiomeric *threo* isomers.

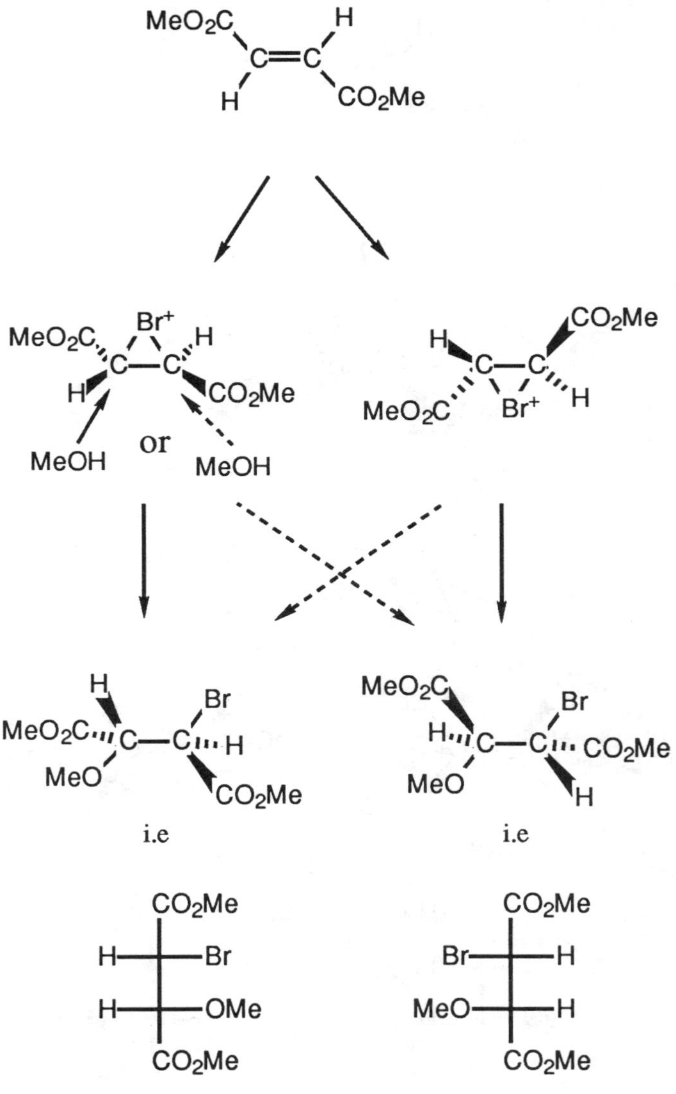

i.e

i.e

$$
\begin{array}{c}
CO_2Me \\
H \underset{}{\overline{}} Br \\
H \underset{}{\overline{}} OMe \\
CO_2Me
\end{array}
\qquad
\begin{array}{c}
CO_2Me \\
Br \underset{}{\overline{}} H \\
MeO \underset{}{\overline{}} H \\
CO_2Me
\end{array}
$$

(1)   (4)

erythro

i.e

i.e

(3)

(2)

threo

36

DISK 4: CONFORMATIONS

4.1 Conformations of Alkanes and Substituted Alkanes

On Disks 2 and 3 there have been several examples of open chain carbon compounds in **eclipsed** and **staggered** arrangements. These so-called **CONFORMATIONS** are illustrated below for bromoethane.

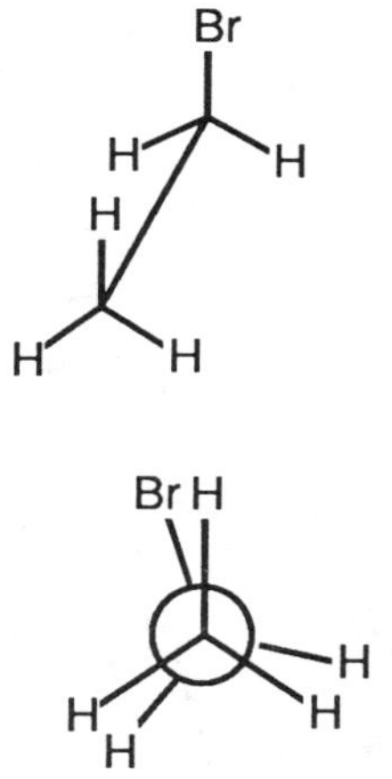

eclipsed 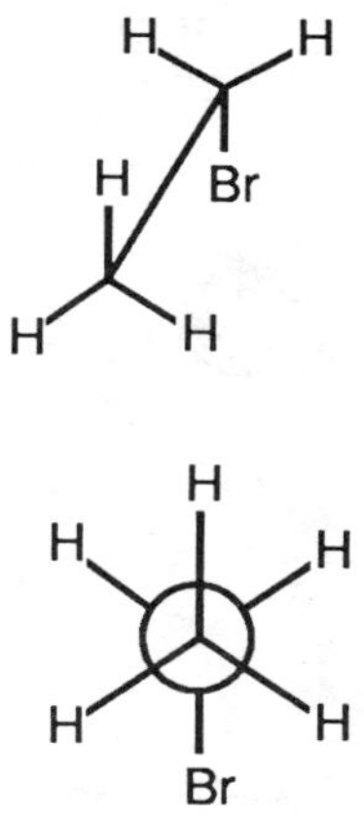   staggered

These conformations are rapidly interconverted at room temperature (we shall return to this point shortly) but **they are not equal in energy.** The staggered conformation is lower in energy than the eclipsed conformation because it minimizes the strain produced by through space interactions between substituents. In the eclipsed conformation, the bonds bearing substituents on adjacent carbon atoms are aligned and the steric interaction between substituents is at a maximum.

The sequence on Disk 4 shows how the energy of ethane varies as one methyl group is rotated relative to the other. The energy curve is reproduced below. The TORSION ANGLE α is the angle between the C-H bonds on adjacent carbon atoms. The energy difference between the staggered and eclipsed conformations can be determined experimentally (by using microwave spectroscopy or thermochemical data). For ethane the value is about 12 kJ mol^{-1}. The origin of the energy barrier in ethane is thought to be a Coulombic interaction between the electrons in the CH bonds which is at a maximum in the eclipsed conformation. The barrier is greater in substituted ethanes and is generally increased by bulky substituents as steric factors make a greater contribution.

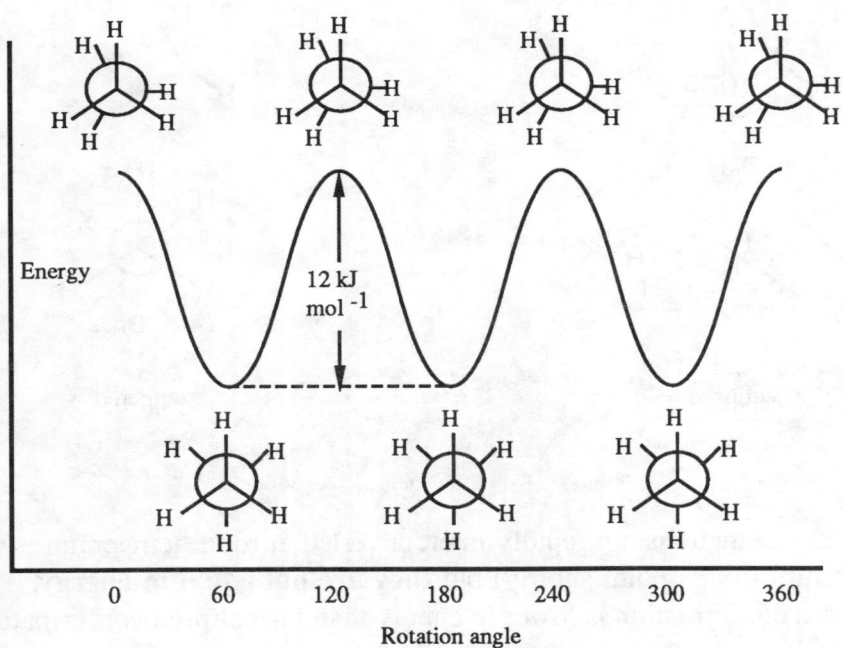

Exercise 4.1. How would you expect the energy curve for propane to differ (if at all) from that for ethane?

At room temperature individual molecules of ethane may have the eclipsed, staggered, or intermediate conformations, but redistribution of energy through collisions is rapid and the rotational energy barrier to interconversion is therefore easily surmounted. For this reason we cannot physically distinguish between ethane molecules in different conformations at room temperature (such conformational isomers are often referred to as **CONFORMERS or ROTAMERS**). As the temperature is lowered the average kinetic energy of the molecules decreases, energy transfer is less, and the energy barrier becomes more significant. There are many instances in which different conformational isomers have been detected and analysed by the use of n.m.r. spectroscopy on samples at low temperatures. The energy barrier is too low for this to be possible for ethane or for bromoethane, but it can be done for cyclohexane, as we shall see in Section 4.2.

The energy curve for rotation about the central carbon-carbon bond of butane is also illustrated on Disk 4. With this molecule there is a new factor to be considered: not all the eclipsed conformations and not all the staggered conformations are of equal energy. The reason is that some of the interactions are between methyl groups on adjacent carbon atoms whereas others are between hydrogens or between methyl groups and hydrogens. The energy curve is reproduced below. The conformation of highest energy is that in which the two methyl groups are aligned and the one of lowest energy is that in which the methyl groups are at $180°$ to each other. This rotamer of minimum energy is called the *ANTI* conformation. Two other eclipsed conformations exist at torsion angles of $120°$ and $240°$. In these two conformations the methyl groups are co-linear with hydrogen atoms. There are also two energy minima on the curve at torsion angles of $60°$ and $300°$. These two staggered structures are called *GAUCHE* conformations in order to distinguish them from the staggered conformation of lowest energy (the *anti* form). Note that the energy difference between the most stable and the least stable rotamers of butane is greater than in ethane: about 19 kJ mol^{-1}.

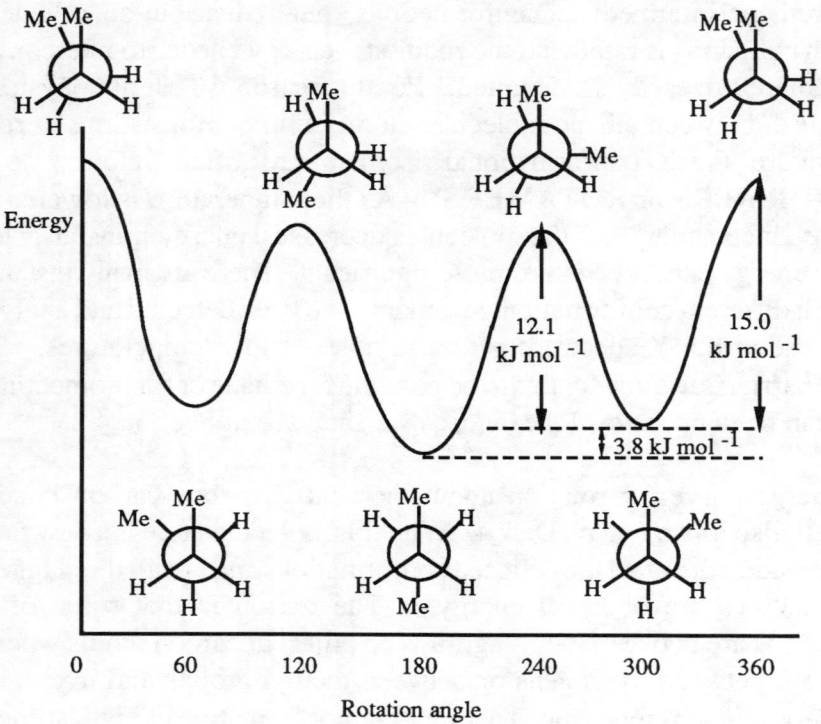

Exercise 4.2. The energy curve for 2-methylbutane is shown on Disk 4 and is reproduced on the opposite page. As an extension to the problem on Disk 4, draw structures for the conformations corresponding to each of the maxima and each of the minima on the curve (use Newman projection formulae).

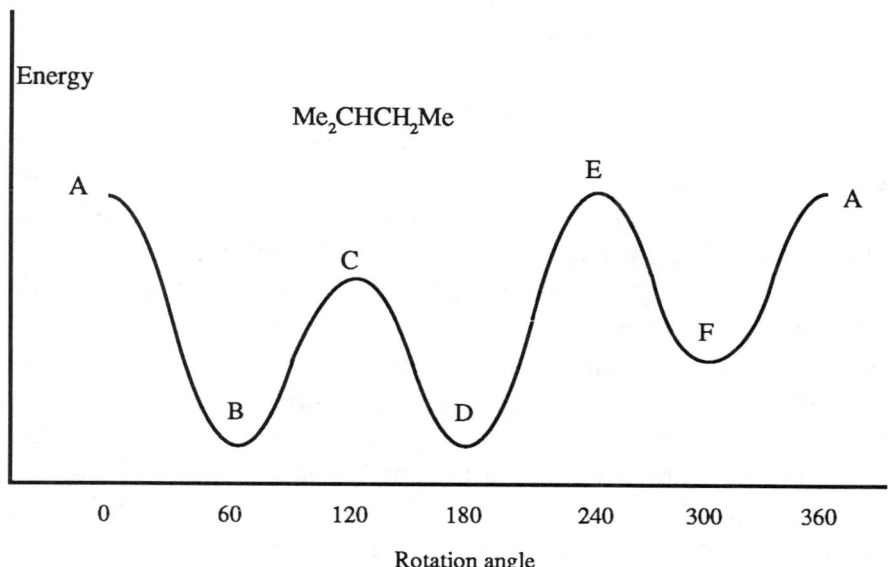

Energy

Me$_2$CHCH$_2$Me

Rotation angle

Where there is more than one conformation of minimum energy, as in butane, the relative population of the conformers can be calculated from the Boltzmann distribution

$$N_i / N_j \quad = \quad \exp [- \Delta E / kT]$$

where N_i and N_j are the number of molecules in states i and j respectively, ΔE is the energy difference between the two states ($E_i - E_j$), k is the Boltzmann constant (1.381×10^{-23} J K^{-1}) and T the temperature.

Exercise 4.3. Calculate the relative populations of the *gauche* and *anti* conformers in butane at 298K using the Boltzmann equation and the data in the figure. Note that there are two *gauche* conformers with the same energy.

4.2 Conformations of Cyclohexane and Substituted Cyclohexanes

Substituted cyclohexanes are very common structural units in organic chemistry and their conformations are of particular importance. As with simple alkanes, the preferred conformations of cyclohexanes are those which minimize through space steric interactions. A planar structure would be far too strained: the hydrogen atoms on each side of the ring would be eclipsed and the bond angles would be strained ($120°$ instead of the optimum angle of $109°$). The preferred conformation of cyclohexane is a staggered structure called a **CHAIR** structure. For cyclohexane there are two chair conformations of equal energy; at room temperature the two forms are in rapid equilibrium, as indicated below. The Newman projection formulae shown below the structures indicate the conformations about one of the carbon- carbon bonds. From these formulae it is clear that the chair conformations are staggered structures with a *gauche* interaction between the CH_2 groups.

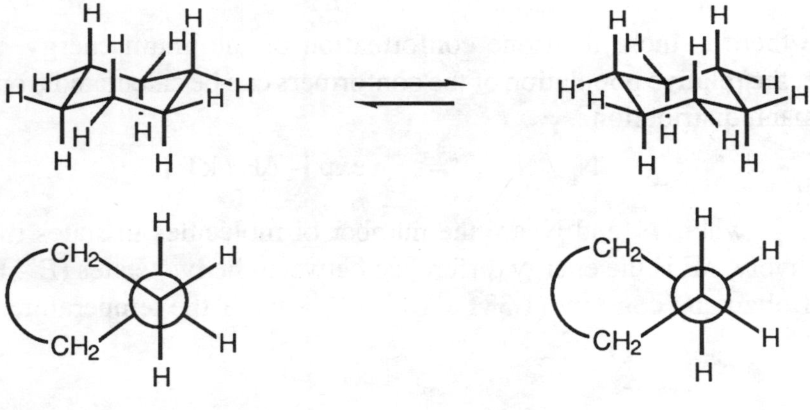

The chair conformers contain two different types of hydrogen substituent: six attached to the bonds in the vertical plane (as the structure is drawn here) called **AXIAL** hydrogens, and six attached to the bonds in the horizontal plane, called **EQUATORIAL** hydrogens. The environment of the two types is different and we might expect them to have different chemical shifts in n.m.r. spectra, for example. At room temperature, this is not the case because the conformer is converted too rapidly into its other chair form. In this interconversion, or "flipping", the axial substituents of one conformer become the equatorial substituents of the other, and vice-versa. The interconversion is slower the lower the temperature, and it is possible to see separate signals for equatorial and axial hydrogens in the n.m.r. spectrum of cyclohexane if the spectrum is recorded at low temperature (at about -80°C or below).

You will need to be able to draw the chair form of cyclohexane and to place the equatorial and axial bonds properly. One method of drawing cyclohexane is as follows:

(i) Draw two parallel lines AB and CD of equal length and in the orientation shown.

(ii) Draw two attached parallel lines AE and DF such that points A and F, and points E and D, are respectively in the same horizontal planes.

(iii) Complete the ring by joining E with C and B with F. Lines EC and BF should then be parallel and equal in length.

(iv) Draw equatorial bonds parallel to the ring bonds as follows:

from A parallel to CE
from B parallel to DF
from C parallel to AE
from D parallel to BF
from E parallel to BA
from F parallel to CD.

(v) Draw axial bonds vertically as follows:

from A, C, and F vertically upwards
from E, B, and D vertically downwards.

Steps (i) to (v) are shown in sequence overleaf.

Stereochemistry

(i)   (ii)   (iii)

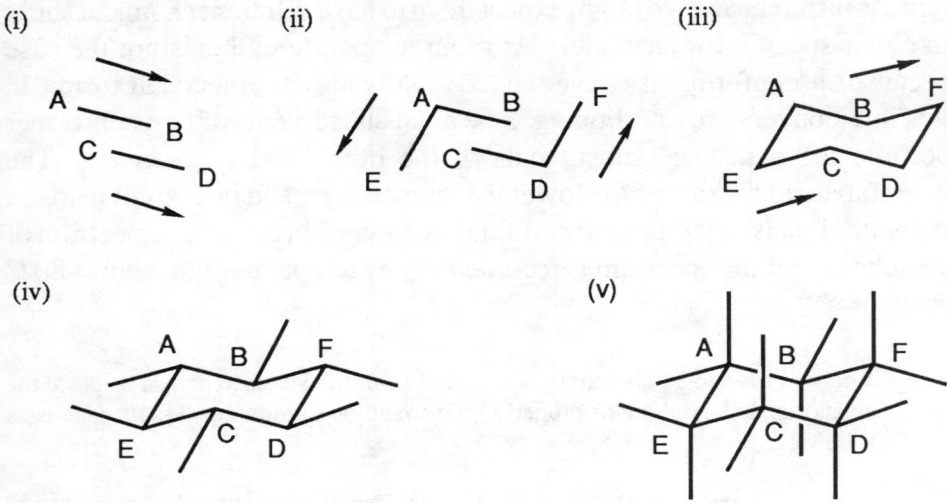

(iv)   (v)

The chair structures are not completely free from strain: there are *gauche* interactions between the CH_2 groups and also interactions between the axial hydrogen atoms.   These repulsive interactions are called **1,3-diaxial interactions**, and are shown below.   1,3-Interactions between equatorial hydrogens are negligible by comparison.

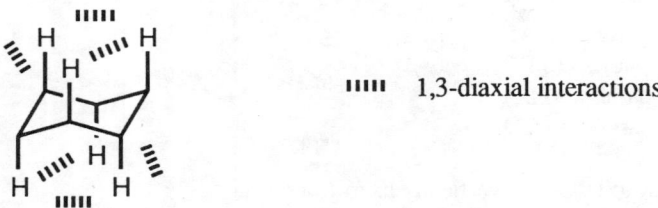

IIIII   1,3-diaxial interactions

When the cyclohexane ring is monosubstituted the two chair conformations are no longer equal in energy. This is illustrated for methylcyclohexane, one chair conformation having the methyl group in an axial position and the other having it in an equatorial position. The lower energy conformation is that with the methyl group equatorial because in the axially substituted conformer there are unfavourable 1,3-diaxial interactions between the methyl group and the axial hydrogens. The equilibrium mixture of chair conformers contains a higher proportion of the lower energy conformer than of the higher energy form.

It is possible to express the preference for equatorially substituted cyclohexanes in a quantitative way by determining the free energy difference $\Delta G°$ between the equatorial and axial conformers. This can be done, for example, by using low temperature n.m.r. spectroscopy. The values for some monosubstituted cyclohexanes are shown in the Table. The greater the free energy difference, the more the equatorial conformer is favoured over the axial conformer.

Table. Free energy values ($\Delta G°$) for the conversion of some monosubstituted cyclohexanes from an axial to an equatorial conformation.

Substituent	$\Delta G°$ (kJ mol^{-1})
F	1.1
Cl	2.2
Br	2.3
I	1.9
Me	7.1
Et	7.3
CMe$_3$	20 (approx.)
Ph	12.5
CO$_2$H	5.9
OH	3.9

With two substituents attached to the cyclohexane ring there are two possible diastereomers, labelled *cis* and *trans*, for each of the positional isomers. These are illustrated below. One way of deciding whether a disubstituted cyclohexane is the *cis* or the *trans* isomer is to imagine a horizontal plane drawn through the ring and bisecting the ring bonds. If the isomer is *cis*-disubstituted the two substituent groups are on the same side of this plane; if it is *trans*-disubstituted they are on opposite sides.

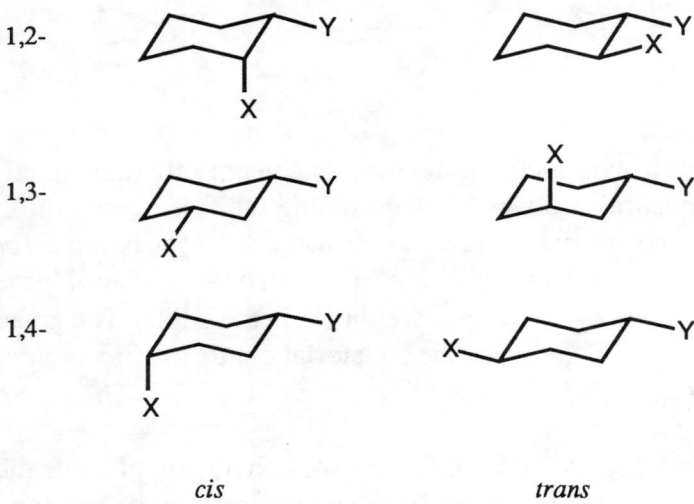

1,2-

1,3-

1,4-

cis   *trans*

Exercise 4.4. Draw structures for each of the following: (a) *cis*- 1,4-dimethylcyclohexane, (b) *trans*-3-bromo-1-chlorocyclohexane, and (c) *trans*-1-chloro-2-methylcyclohexane.

Conformations with two substituents equatorial are clearly of lower energy than those with the two substituents axial: the free energy difference between a diequatorial and a diaxial conformation can be estimated by adding the contributions from the individual substituents listed in the Table above. When the diastereomer has one equatorial and one axial substituent, the substituent with the lower free energy difference in the Table preferentially occupies the axial position. For example, the preferred conformation of *cis*-1-fluoro-4-methylcyclohexane is that which has the fluorine axial and the methyl group equatorial.

Exercise 4.5. Draw each of the following in the preferred conformations: (a) *trans*-1-chloro-3-phenylcyclohexane, (b) *cis*-2-methylcyclohexanol, and (c) *cis*-4-t-butylcyclohexanecarboxylic acid.

Exercise 4.6. Draw chair structures for menthol (A) and for neomenthol (B).

The preference for conformations with bulky substituents in equatorial positions can also be rationalised by counting the number of *gauche* interactions in the equatorial and axial conformers. This is illustrated on Disk 4 for *trans*-1,2-dimethylcyclohexane, and the conformations are also shown overleaf.

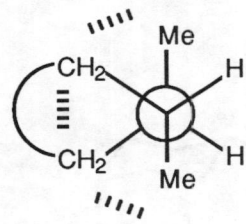

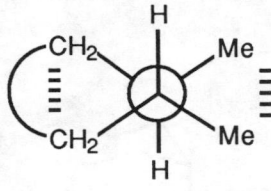

diaxial

three gauche interactions

diequatorial

two gauche interactions

Chair conformations are not the only ones possible for cyclohexanes. Two others are shown below: the **boat** and the **twist boat**, or **skew**. The boat form is about 27 kJ mol^{-1} higher in energy than the chair form: you may notice that it has eclipsing interactions between two pairs of CH_2 groups and a steric interaction between hydrogens on the other two carbon atoms. The boat form is therefore not a stable form of simple cyclohexanes, although it is sometimes found when cyclohexanes are incorporated into polycyclic molecules. The twist boat, or skew, form is derived from the boat by twisting two of the CH_2 groups out of the horizontal plane (as drawn). It is slightly more stable (by about 7 kJ mol^{-1}) than the boat form because some of the eclipsing interactions are reduced. This form is also rarely an important one for simple cyclohexanes. From the Table above you will see that t-butyl groups have a very strong preference for equatorial positions in chair conformations. In the case of a substance such as *cis*-1,4-di-t- butylcyclo-hexane one of the substituents would be forced to occupy an axial position in a chair conformation. The molecule instead adopts a twist boat conformation, which allows both of the bulky substituents to occupy pseudo-equatorial positions.

boat   twist boat (skew)   *cis*-1,4-di-t-butylcyclohexane

The final topic on Disk 4 is the structure of molecules with two fused cyclohexane rings, the decalins. The decalin structural unit is very common in natural products, especially the *trans*-fused structure. The two decalin structures are illustrated below. (The name *trans*-decalin is an indication of the fact that at the ring junction the bonds to the second ring are connected as *trans* substituents; for *cis*-decalin they are connected as *cis* substituents).

trans-decalin   cis-decalin

Exercise 4.7. Construct molecular models of *cis*- and *trans*-decalin. Compare the two and note the rigidity of *trans*-decalin relative to the *cis* form. You should be able to flip the *cis*-decalin model to another structure, also with two chair cyclohexane rings, but this is not possible for the *trans* structure. What unfavourable steric interactions can you identify in the *cis* structure which makes it less stable than the *trans* diastereomer?

Exercise 4.8. Draw chair structures for the most stable forms of each of the following: (a) 2-methyl-*trans*-decalin, (b) 1-chloro-*cis*-decalin, and (c) *cis*-2,3-dibromo-*trans*-decalin.

The *trans* ring fusion can be extended beyond two cyclohexane rings, and indeed there are many polycyclic natural products which have this type of ring fusion. The steroids are a large group of natural products with a tetracyclic ring system made up of three six-membered rings and one five-membered ring linked in the manner shown overleaf. Most of these compounds with fully saturated skeletons have *trans*-fused rings and their structures can be represented as shown in the right hand figure. Androsterone, one of the natural products illustrated on Disk 3, has a similar structure.

The conformational preferences of other cyclic alkanes are not mentioned on Disk 4 but they will be discussed briefly here. The three membered cyclic alkane, cyclopropane, has very little flexibility and the C-H bonds on adjacent carbon atoms are forced into eclipsing interactions. Cyclopropane is an unusually reactive hydrocarbon. The four-membered cyclic alkane, cyclobutane, is not planar. Its structure can be represented as shown below, as two boat-like conformers with a low barrier to interconversion. The molecule is still strained because the bond angles are less than the tetrahedral value and there are 1,3- interactions of C-H bonds.

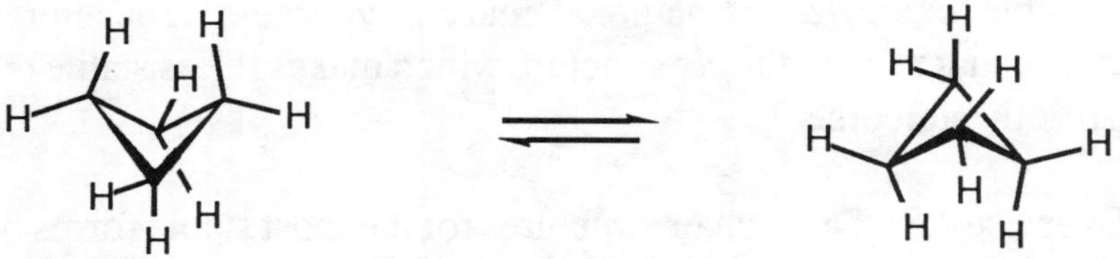

A planar cyclopentane would have little bond angle strain (the bond angle would be 108°) but cyclopentane does not adopt a planar structure because of the eclipsing interactions between adjacent C-H bonds which such a structure would have. Instead it exists as a mixture of rapidly interconverting structures, with no definite energy minimum, produced by up and down motion of the methylene groups.   Two such types of nonplanar cyclopentane, the "envelope" and the "half chair" forms, are illustrated below. The up and down motion of the methylene groups in effect "rotates" about the ring. This type of interconversion is sometimes referred to as **PSEUDOROTATION**.

envelope conformers   half chair conformers

Cycloalkanes having a ring size greater than six might be expected to be free from strain, but this is not the case. Indeed the strain energy increases up to a ring size of nine or ten and then decreases: only rings larger than this are relatively free from strain. The reason is illustrated by the conformation of cyclodecane shown below. The rings must be puckered in order to achieve the normal tetrahedral bond angles but in the medium sized rings this causes interactions between hydrogen atoms on opposite sides of the rings (called transannular interactions). The rings are conformationally mobile but none of the conformations allows these transannular interactions to be avoided.

transannular interactions

This completes our coverage of the four Disks. If you have followed the material and mastered the jargon you will be in a good position to go on to more advanced stereochemical topics, especially to understand how stereochemistry influences the outcome of chemical reactions.

ANSWERS TO EXERCISES

1.2   (a) no; (b) yes; (c) no; (d); yes.

1.3   (a)

(c)

1.4   (a) yes; (b) yes; (c) no; (d) yes; (e) no; (f) yes; (g) no.

1.5   +42.5

1.6   (a) 80%; (b) 9 : 1

1.7   (a) OH; (b) $CHCl_2$; (c) CH_2NH_2; (d) $CONH_2$; (e) $CH=CH_2$.

1.8   (a) F, CH_2D, CH_3, Li

(b) CHO, CN, C_6H_5, $CH=CH_2$

(c) $C\equiv CH$, $CH=CH_2$, CH_2CH_2OH, CH_2CF_3

1.9   (a) S; (b) R; (c) S; (d) R; (e) S; (f) S; (g) R.

1.10   (a)   (b)   (c)   (d)

2.1   (a)   (b)

2.2 (a)

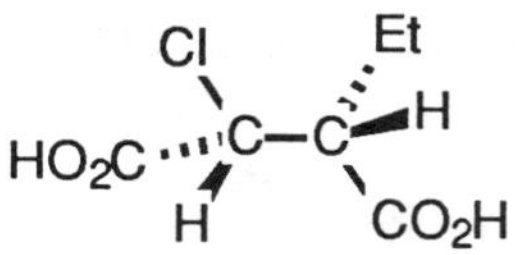

Cl   Et
    | |
HO₂C⋯C—C⟍H
    | |
    H   CO₂H

(b)

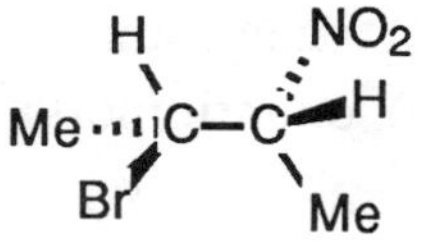

    H   NO₂
    | |
Me⋯C—C⟍H
    | |
    Br   Me

2.3 (a)

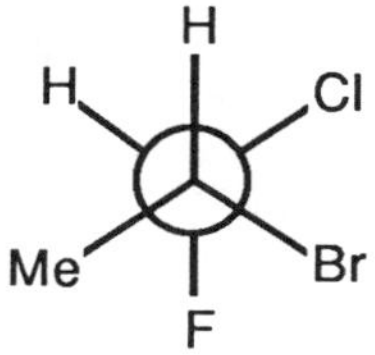

    H
H   Cl
   ( )
Me   Br
    F

(b)

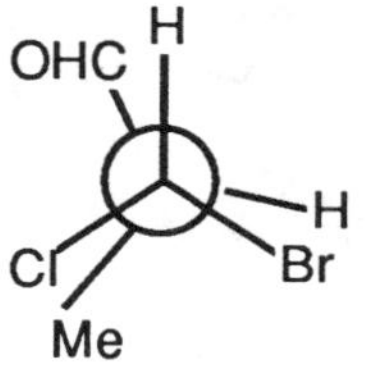

OHC H
   ( )
      H
Cl   Br
   Me

(c)

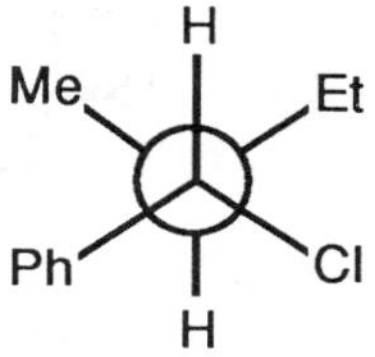

    H
Me   Et
   ( )
Ph   Cl
    H

2.4 (a) *R*; (b) *S*.

2.6

CHO
H——Cl   ↷   Me
Me   Cl——H   i. e. 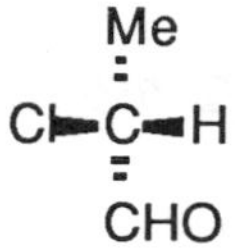   Me   identical to
                CHO   Cl—C—H
                                  CHO   original.

2.7 (a)

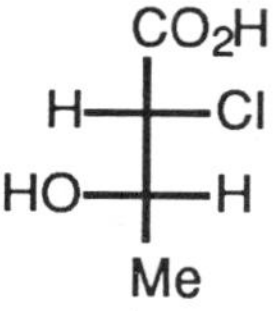

CO₂H
H——Cl
HO——H
Me

(b)

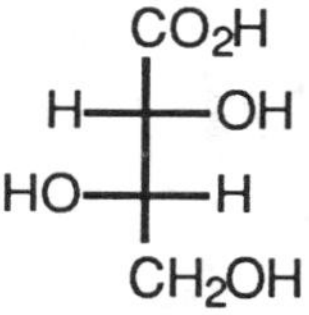

CO₂H
H——OH
HO——H
CH₂OH

(c)

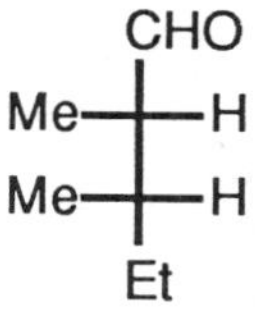

CHO
Me——H
Me——H
Et

(d)

CO₂H
MeCONH——H
HO——H
Me

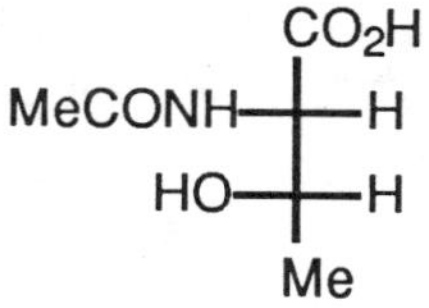

2.8 (a) 1*S*, 2*R*; (b) 1*S*, 2*S*.

3.2 (iii) and (iv) enantiomers;

(i) and (iv), (ii) and (iii), (ii) and (iv) diastereomers.

Stereochemistry

3.3   (a) identical; (b) diastereomeric; (c) identical;
(d) enantiomeric; (e) diastereomeric; (f) diastereomeric;
(g) enantiomeric.

3.5   (a)

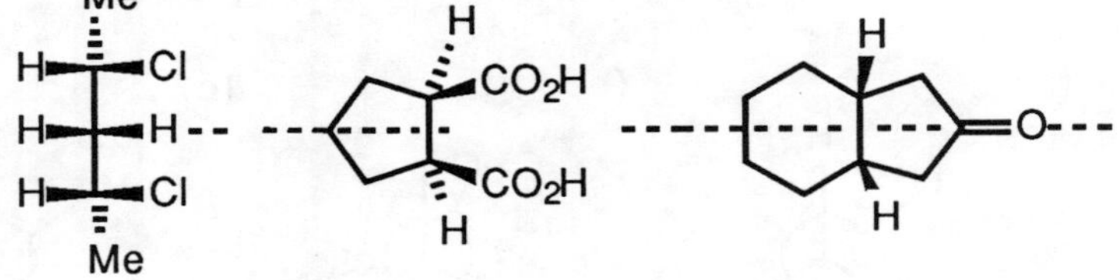

3.6

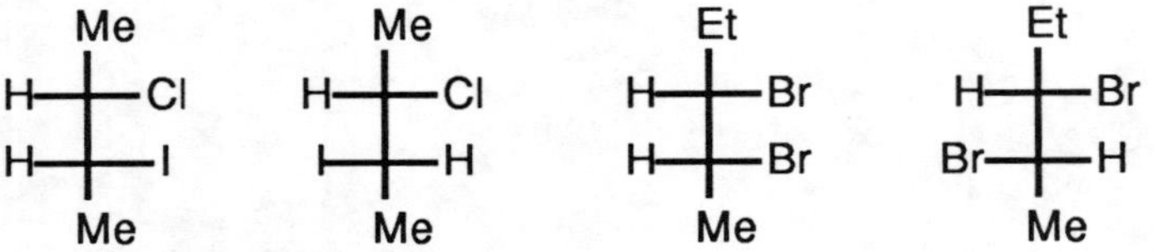

erythro   *threo*   *erythro*   *threo*

4.1   Same shape; energy difference between maxima and minima
greater because of H--Me interactions in eclipsed conformations.

4.2

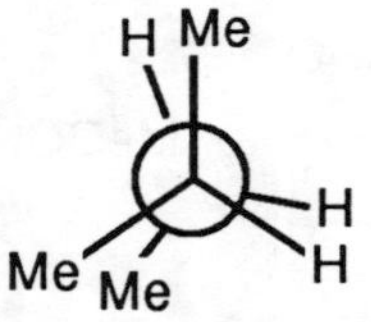

A 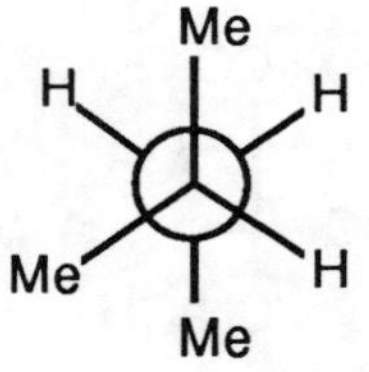   B 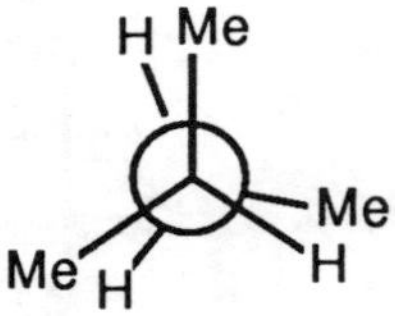   C

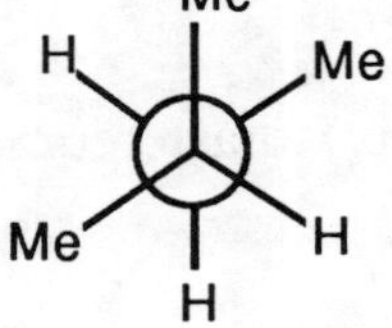

D 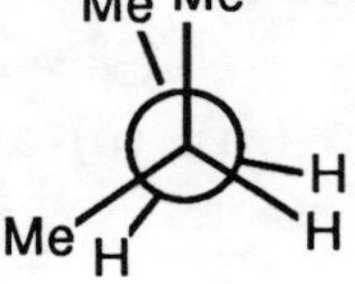   E 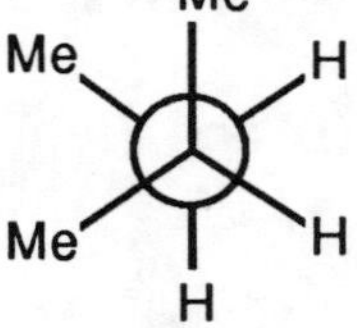   F

4.3 N_{gauche} / N_{anti} = 0.414 (i.e. 2 x 0.207)

4.4 (a)   (b)   (c)

4.5 (a)   (b)   (c)

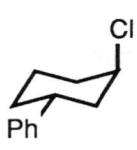

4.6

A

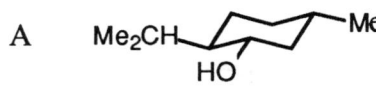

B

4.8 (a)   (b)   (c)

Stereochemistry

INDEX